The Deep Sky Observer's Guide

Astronomical Observing Lists Detailing Over 1,300 Night
Sky Objects for Binoculars and Small Telescopes

For all the amateur astronomers around the world

who only wish for a tall telescope

and a star to steer her by

Contents

Introduction

Why I Wrote This Book

The idea for this book really came about as a solution to a problem. I was conducting research on deep sky objects for books and had to manually draw up a list of objects to observe each night.

My view of the sky is limited to the western hemisphere; I can't see the east at all. I therefore needed to know which objects were well placed and which were sinking in the west and on the "last chance to see" list. Manually creating the list wasn't a huge problem, but it was a little time consuming as I had to consider the positions of the objects in question.

I also wanted a list of highlighted objects to observe once I'd worked my way through my targets for the night. I didn't want to rely on electronics (such as cell phone apps or other software) that would either fail me as the batteries died or ruin my night vision.

It would be nice, I thought, to have this all pre-prepared and have it available to me, ready to grab 'n go in my eyepiece box.

In short, I wanted a book that could tell me what objects could be seen at almost any time of night, from any part of the world, at any time of year. I wanted to know roughly what part of the sky they might be in and I wanted every object to be visible in a small scope like mine. And it had to be conveniently pocket-sized.

As always, if I can't find the book I want, I'll write it myself.

My goal then was to create something that anyone with some experience could use, whether they had a GoTo 'scope or wanted to hunt down some DSO's with their trusty sky chart.

Consequently, I spent months compiling twenty-four lists of objects that should be visible with a small telescope (75mm or larger.) My criteria for selecting the objects was as follows:

- Double stars with at least 2" separation. The primary should be no fainter than magnitude 6.0 with the secondary being no fainter than magnitude 8.5
- Open clusters brighter than magnitude 9.0
- Globular clusters brighter than magnitude 10.0
- Nebulae under magnitude 10.0

- Planetary nebulae under magnitude 12.0
- Galaxies under magnitude 12.0

When I was done, I found myself with some 1,300 objects – far more than I'd anticipated and many of them potentially visible with binoculars too.

How to Use this Book

There's basically two ways you can use this book. If you're looking to spend a short time outside (or only have a short time left at the end of your observing session) you can use the *Quick Observing Lists* for 10 to 12 suggested objects to observe.

If, on the other hand, you're looking to go deeper, you can use the *Detailed Lists* that each cover an hour of right ascension and contain, on average about fifty objects (both northern and southern hemispheres.)

Although everything listed can, theoretically, be observed with a small telescope, an objects' visibility will greatly depend upon your location in two important ways:

1. For most of the fainter objects (specifically, those fainter than magnitude 9, such as many planetary nebulae and galaxies) you'll need clear, dark skies to see them properly. You may well be able to see them from the suburbs of a small town, but you probably won't see much. If you're observing from a large town or city, you'll be lucky to see them at all. (Multiple stars and bright open clusters are usually no problem at all.)

2. Which objects you can see will also depend upon your location's latitude. If, for example, you live in England, you won't be able to see many of the objects in the southern hemisphere. See the *Latitude and Declination Table* in the Appendix for more information.

To Use the Quick Observing Lists

1. Look for your current time of year in the table. The early part of the month is defined as being the first half (ie, from the 1st to the 15th) while the late part of the month is the second half (from the 16th to the end of the month.)

2. If your location is observing daylight savings time, you'll first need to subtract this from your current time before looking up the chart #. For example, if you live in the US or the UK, summer time puts the clocks one hour ahead. So if it's currently 11pm, you'll need to look for 10pm in the table.

3. Look for the appropriate chart # in the table and turn to that section in the book.

4. You'll see two charts – one for the northern and one for the southern hemisphere.

5. On the left, you'll see a table of *Detailed Lists* that will allow you to observe many more objects. The lists on the left are rising in the east while those on the right are setting on the west.

6. Also on the left, you'll see a table detailing the number of different DSO's that are currently visible, directly underneath the corresponding list number. So if you're specifically looking to observe open clusters, you'll know how many to look for and what area of the sky to concentrate your search.

7. On the right, you'll see a table of suggested objects to observe. I've tried to suggest objects that should be visible for the majority of readers (so I'll admit to a bias toward the northern hemisphere). I've also tried to make each table a little different; although there is a little repetition, I've tried to avoid listing the same object on consecutive tables. That way, if you're outside for a few hours, there should be some variety.

To Use the Detailed Lists

1. As with the Quick Observing Lists, you'll need to look for the table detailing your current time of year.

2. You'll also need to deduct daylight savings time before referring to the table. (For example, if your area adds an hour for summer time and it's 11pm, you'll need to look for 10pm in the table.)

3. You'll see that the available lists are organized from east on the left, through to the zenith in the middle and west on the right. Those objects that are low over the eastern horizon are on the far left and those that are low on the western horizon are on the far right.

4. Turn to the specific list to begin observing. Objects are listed in the following order:

    *   Multiple stars
    *   Open Clusters
    *   Open Clusters with Nebulae
    *   Nebulae
    *   Planetary Nebulae
    *   Globular Clusters
    *   Galaxies

5. You'll also see images (taken with *Slooh*) for some of the objects listed.

About the Images

The star charts were created using the *Mobile Observatory* app. Written by Wolfgang Zima, it's available for Android devices and is available through the Google Play store.

The object images were all taken using *Slooh*, an online tool that allows users to photograph the night sky using observatories in the Canary Islands and Chile. You can learn more at http://www.slooh.com

One Last Thing

Unfortunately, I'm not able to provide locator star charts for each of the objects (time, formatting and printing costs make it difficult) but the R.A. and Declination co-ordinates should be enough to locate the object on a popular star chart, such as the *Pocket Sky Atlas*.

Of course, if you own a GoTo 'scope, you'll be able to search for many within the telescope's built database or enter the co-ordinates for any that have not already been pre-programmed.

I also have no real control over the number of objects that are visible during any particular time. I could, for example, make each list a specific size and set a target of, say, fifteen of each object – but then I'd either be leaving some easily observable objects off the list or adding a lot of "filler" objects that might be beyond the range of small 'scopes.

Instead, I thought it best to stick with the criteria I'd originally set and allow myself the challenge of observing something new!

Also by the Author...

2016 An Astronomical Year is written for everyone with an interest in astronomy and contains information on hundreds of night sky events throughout the year. It was designed for astronomers of all levels and includes details of the lunar phases and eclipses, as well as conjunctions, oppositions, magnitude and apparent diameter changes for the planets and major asteroids.

To date, the 2015 edition has been downloaded nearly 3,000 times, was ranked #1 in Free Kindle Astronomy books, within the Top 10 Paid Kindle Astronomy books and within the Top 50 Free Kindle Non-Fiction books.

It is available in paperback and Kindle editions in the United States, Canada and the United Kingdom. (Please be aware that due to the cost of printing in color, the paperback does not contain images and is purely text only.)

2016 The Night Sky Sights is specifically designed for absolute beginners and casual stargazers without a telescope. The guide highlights over 125 astronomical events in 2016 - all of them visible with just your eyes - and showcases events visible in both the evening and pre-dawn sky as well as

those you can see throughout the night.

It is currently available in paperback and Kindle editions in the United States, Canada and the United Kingdom.

The Astronomical Almanac (2016-2020): A Comprehensive Guide to Night Sky Events provides details of thousands of astronomical events from 2016 to the end of 2020. Designed for more experience astronomers, this the guide includes almost daily data and information on the Moon and planets, as well as Pluto, Ceres, Pallas, Juno and Vesta.

To date, the 2015-2019 edition has been downloaded nearly 6,000 times, was ranked #1 in the Free Kindle Astronomy book category, #3 in the Paid Kindle Astronomy book category and within the Top 50 of *all* Free Kindle books in October 2014.

It is available in paperback and Kindle editions worldwide, including the United States, Canada, the United Kingdom and Australia.

The Amateur Astronomer's Notebook: A Journal for Recording and Sketching Astronomical Observations is the perfect way to log your observations of the Moon, stars, planets and deep sky objects. It is available as both a full-size 8.5" by 11" journal and also as a 5" by 8" pocket notebook. The larger edition has room for 150 observing sessions while the pocket edition allows you to record 100 observations.

It is available as a paperback in selected areas. (Full Size Edition: United States, Canada and the United Kingdom. Pocket Edition: United States, Canada and the United Kingdom.)

Echoes of Earth – a collection of science fiction, mythological and philosophical short stories that I wrote many, many moons ago. (i.e., in the mid 1990's.)

It is available as a Kindle edition in selected areas. (United States, Canada, the United Kingdom and Australia.)

The Author Online

Email: astronomywriter@gmail.com

Amazon US: http://tinyurl.com/rjbamazon-us

Amazon UK: http://tinyurl.com/rjbamazon-uk

The Astronomical Year: http://tinyurl.com/theastroyear

Facebook: http://tinyurl.com/rjbfacebook

Twitter (@astronomywriter): http://tinyurl.com/rjbtwitter

Clear skies,

Richard J. Bartlett

September 27th, 2015

The Quick Observing Lists

Evening Hours

If observing during daylight savings time, first deduct one hour and then refer to the corresponding chart number for observable deep sky objects.

	6pm	7pm	8pm	9pm	10pm	11pm
Early January	1	2	3	4	5	6
Late January	2	3	4	5	6	7
Early February	3	4	5	6	7	8
Late February	4	5	6	7	8	9
Early March	5	6	7	8	9	10
Late March	6	7	8	9	10	11
Early April	7	8	9	10	11	12
Late April	8	9	10	11	12	13
Early May	9	10	11	12	13	14
Late May	10	11	12	13	14	15
Early June	11	12	13	14	15	16
Late June	12	13	14	15	16	17
Early July	13	14	15	16	17	18
Late July	14	15	16	17	18	19
Early August	15	16	17	18	19	20
Late August	16	17	18	19	20	21
Early September	17	18	19	20	21	22
Late September	18	19	20	21	22	23
Early October	19	20	21	22	23	24
Late October	20	21	22	23	24	1
Early November	21	22	23	24	1	2
Late November	22	23	24	1	2	3
Early December	23	24	1	2	3	4
Late December	24	1	2	3	4	5

Pre-Dawn Hours

If observing during daylight savings time, first deduct one hour and then refer to the corresponding chart number for observable deep sky objects.

	12am	1am	2am	3am	4am	5am
Early January	7	8	9	10	11	12
Late January	8	9	10	11	12	13
Early February	9	10	11	12	13	14
Late February	10	11	12	13	14	15
Early March	11	12	13	14	15	16
Late March	12	13	14	15	16	17
Early April	13	14	15	16	17	18
Late April	14	15	16	17	18	19
Early May	15	16	17	18	19	20
Late May	16					
Early June	17	18	19	20	21	22
Late June	18	19	20	21	22	23
Early July	19	20	21	22	23	24
Late July	20	21	22	23	24	1
Early August	21	22	23	24	1	2
Late August	22	23	24	1	2	3
Early September	23	24	1	2	3	4
Late September	24	1	2	3	4	5
Early October	1	2	3	4	5	6
Late October	2	3	4	5	6	7
Early November	3	4	5	6	7	8
Late November	4	5	6	7	8	9
Early December	5	6	7	8	9	10
Late December	6	7	8	9	10	11

Chart 1

Northern Hemisphere Sky Chart

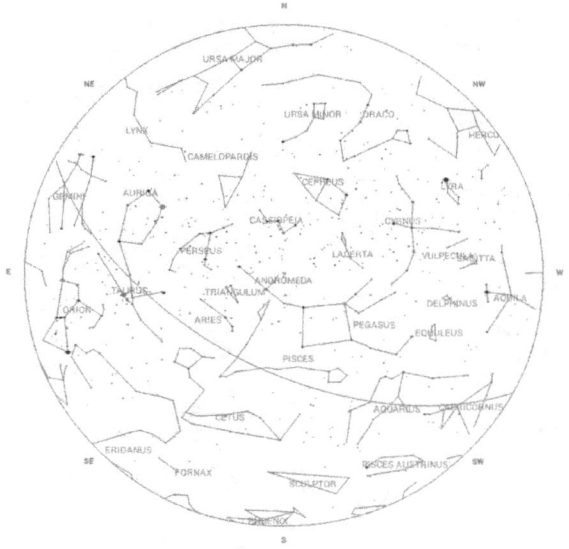

	Eastern Hemisphere				N/S	Western Hemisphere			
	Low			High	Z	High			Low
List #	5	4	3	2	1	24	23	22	21

See the individual lists for more observable objects

Number of DSO's per List

MS	24	19	14	14	15	6	13	15	15
OC	13	8	9	12	6	0	2	7	5
OC/N	1	2	3	0	1	0	3	3	1
Neb	2	1	0	0	1	0	0	0	2
PN	1	3	0	1	0	0	1	2	2
GC	2	0	1	0	2	1	0	2	4
Gx	12	36	27	19	21	11	19	10	5
Total	55	69	54	46	46	18	38	39	34

Chart 1

Southern Hemisphere Sky Chart

Suggested Objects for Both Hemispheres (from west to east)

Con.	Object	Type	R.A.	Dec.	List #
Cyg	M 39	OC	21h 32m	+48° 27'	22
Cas	M 52	OC	23h 25m	+61° 36'	23
	NGC 185	Gx E	00h 39m	+48° 20'	1
	Eta Cas	MS	00h 49m	+57° 49'	1
And	M 31	Gx S	00h 43m	+41° 16'	1
Cet	NGC 247	Gx S	00h 48m	-20° 46'	1
Scl	NGC 300	Gx S	00h 55m	-37° 41'	1
Cas	M 103	OC	01h 33m	+60° 39'	2
	NGC 663	OC	01h 46m	+61° 13'	2
And	59 And	MS	02h 11m	+39° 02'	2
Per	M 34	OC	02h 47m	+42° 45'	3
	STF 331	MS	03h 01m	+52° 21'	3
	Eps Per	MS	03h 58m	+40° 01'	4

Chart 2

Northern Hemisphere Sky Chart

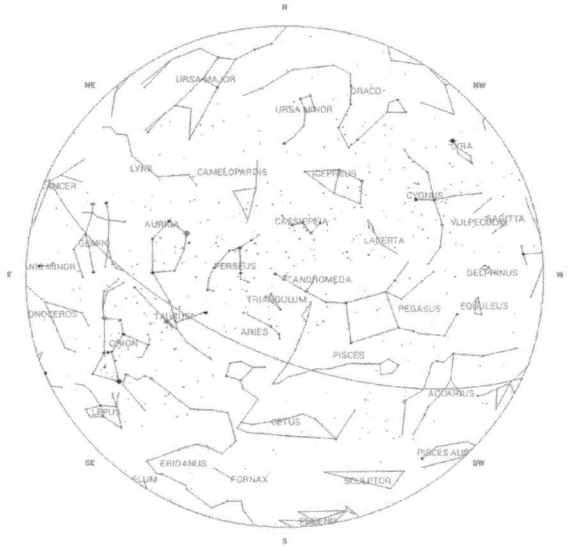

	Eastern Hemisphere			N/S	Western Hemisphere			
	Low		High	Z	High			Low
List #	6	5	4	3	2	1	24	23	22

See the individual lists for more observable objects

Number of DSO's per List

MS	25	24	19	14	14	15	6	13	15
OC	13	13	8	9	12	6	0	2	7
OC/N	7	1	2	3	0	1	0	3	3
Neb	7	2	1	0	0	1	0	0	0
PN	2	1	3	0	1	0	0	1	2
GC	0	2	0	1	0	2	1	0	2
Gx	8	12	36	27	19	21	11	19	10

Total	62	55	69	54	46	46	18	38	39

Chart 2

Southern Hemisphere Sky Chart

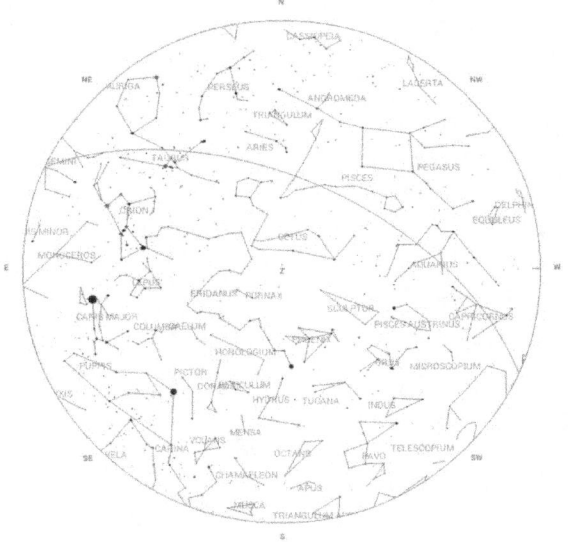

Suggested Objects for Both Hemispheres (from west to east)

Con.	Object	Type	R.A.	Dec.	List #
Lac	8 Lac	MS	22h 36m	+39° 38'	23
Scl	NGC 253	Gx S	00h 48m	-25°17'	1
	NGC 288	OC	00h 52m	-26° 35'	1
Psc	M74	Gx S	01h 37m	+15° 47'	2
Per	M76	PN	01h 42m	+51° 34'	2
Ari	Gam Ari	MS	01h 54m	+19° 18'	2
And	Gam And	MS	02h 04m	+42° 20'	2
Tri	NGC 925	Gx S	02h 27m	+33° 35'	2
	30 Ari	MS	02h 37m	+24° 39'	3
For	NGC 1316	Gx S	03h 23m	-37° 13'	3
Tau	M 45	OC/N	03h 47m	+24° 07'	4
	Mel 25	OC	04h 27m	+16° 00'	4
Cam	1 Cam	MS	04h 32m	+53° 55'	5

Chart 3

Northern Hemisphere Sky Chart

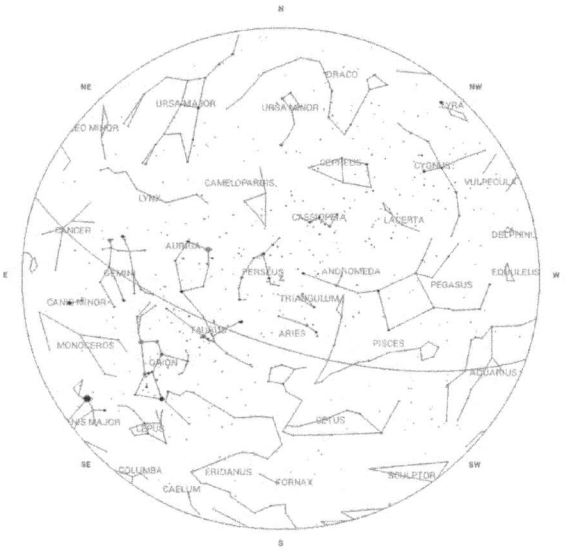

	Eastern Hemisphere				N/S	Western Hemisphere			
	Low			High	Z	High			Low
List #	7	6	5	4	3	2	1	24	23

See the individual lists for more observable objects

Number of DSO's per List

MS	24	25	24	19	14	14	15	6	13
OC	31	13	13	8	9	12	6	0	2
OC/N	6	7	1	2	3	0	1	0	3
Neb	2	7	2	1	0	0	1	0	0
PN	1	2	1	3	0	1	0	0	1
GC	1	0	2	0	1	0	2	1	0
Gx	6	8	12	36	27	19	21	11	19
Total	71	62	55	69	54	46	46	18	38

Chart 3

Southern Hemisphere Sky Chart

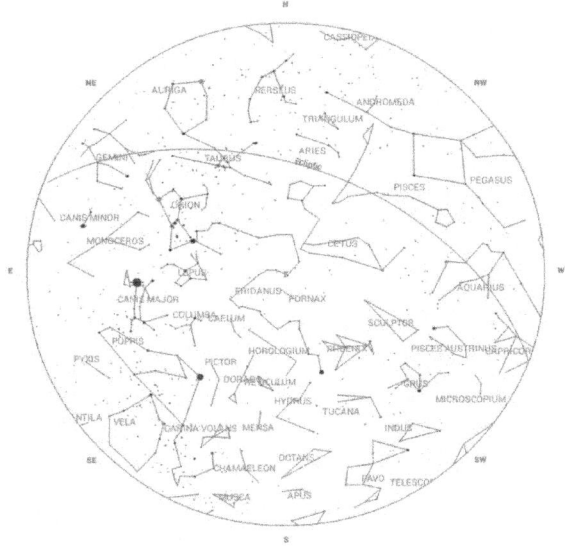

Suggested Objects for Both Hemispheres (from west to east)

Con.	Object	Type	R.A.	Dec.	List #
Cas	NGC 225	OC	00h 44m	+61° 47'	1
	NGC 457	OC	01h 20m	+58° 17'	1
	M 103	OC	01h 33m	+60° 39'	2
	NGC 663	OC	01h 46m	+61° 13'	2
And	59 And	MS	02h 11m	+39° 02'	2
Psc	M 77	Gx S	02h 43m	-00° 01'	3
Eri	NGC 1291	Gx S	03h 17m	-41° 06'	3
	NGC 1535	PN	04h 14m	-12° 44'	4
Per	NGC 1528	OC	04h 16m	+51° 13'	4
Aur	M 38	OC	05h 29m	+35° 51'	5
Ori	Lam Ori	MS	05h 35m	+09° 56'	6
	M 42	Neb	05h 35m	-05° 23'	6
Mon	Bet Mon	MS	06h 29m	-07° 02'	6

Chart 4

Northern Hemisphere Sky Chart

	Eastern Hemisphere			N/S	Western Hemisphere			
	Low			High	Z	High			Low
List #	8	7	6	5	4	3	2	1	24

See the individual lists for more observable objects

	Number of DSO's per List								
MS	20	24	25	24	19	14	14	15	6
OC	27	31	13	13	8	9	12	6	0
OC/N	2	6	7	1	2	3	0	1	0
Neb	0	2	7	2	1	0	0	1	0
PN	0	1	2	1	3	0	1	0	0
GC	1	1	0	2	0	1	0	2	1
Gx	4	6	8	12	36	27	19	21	11
Total	54	71	62	55	69	54	46	46	18

Chart 4

Southern Hemisphere Sky Chart

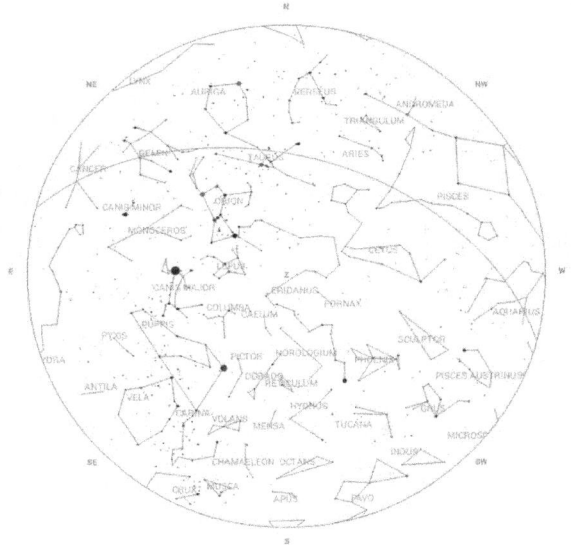

Suggested Objects for Both Hemispheres (from west to east)

Con.	Object	Type	R.A.	Dec.	List #
Psc	Psi[1] Psc	MS	01h 06m	+21° 28'	1
Tri	M 33	Gx S	01h 34m	+30° 40'	2
And	NGC 752	OC	01h 58m	+37° 51'	2
Ari	30 Ari	MS	02h 37m	+24° 39'	3
For	NGC 1316	Gx S	03h 23m	-37° 13'	3
	NGC 1365	Gx S	03h 34m	-36° 08'	4
Cam	IC 342	Gx S	03h 47m	+68° 06'	4
Tau	M 45	OC/N	03h 47m	+24° 07'	4
Col	NGC 1851	GC	05h 14m	-40° 03'	5
Ori	The Ori	MS	05h 35m	-05° 25'	6
	STF 747	MS	05h 35m	-06° 00'	6
Gem	M 35	OC	06h 09m	+24° 21'	6
CMa	M 41	OC	06h 46m	-20° 45'	7

Chart 5

Northern Hemisphere Sky Chart

	Eastern Hemisphere				N/S	Western Hemisphere		
	Low			High	Z	High		Low
List #	9	8	7	6	5	4	3	2	1

See the individual lists for more observable objects

Number of DSO's per List

MS	19	20	24	25	24	19	14	14	15
OC	12	27	31	13	13	8	9	12	6
OC/N	1	2	6	7	1	2	3	0	1
Neb	0	0	2	7	2	1	0	0	1
PN	2	0	1	2	1	3	0	1	0
GC	1	1	1	0	2	0	1	0	2
Gx	13	4	6	8	12	36	27	19	21
Total	48	54	71	62	55	69	54	46	46

Chart 5

Southern Hemisphere Sky Chart

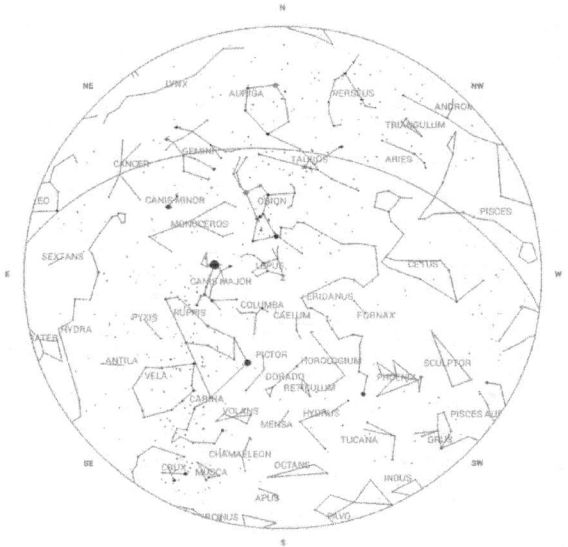

Suggested Objects for Both Hemispheres (from west to east)

Con.	Object	Type	R.A.	Dec.	List #
Ari	Lam Ari	MS	01h 58m	+23° 36'	2
Per	NGC 869/884	OC	02h 19m	+57° 08'	2
	M34	OC	02h 47m	+42° 45'	3
	STF 331	MS	03h 01m	+52° 21'	3
Eri	NGC 1535	Gx	04h 14m	-12° 44'	4
Per	NGC 1528	OC	04h 16m	+51° 13'	4
Lep	M 79	GC	05h 24m	-24° 31'	5
Aur	M 36	OC	05h 36m	+34° 08'	6
	M 37	OC	05h 52m	+32° 33'	6
	NGC 2281	OC	06h 48m	+41° 05'	7
CMa	145 CMa	MS	07h 17m	-23° 19'	7
Pup	M 47	OC	07h 37m	-14° 29'	8
	Kap Pup	MS	07h 39m	-26° 48'	8

Chart 6

Northern Hemisphere Sky Chart

	Eastern Hemisphere			N/S	Western Hemisphere			
	Low			High	Z	High			Low
List #	10	9	8	7	6	5	4	3	2

See the individual lists for more observable objects

Number of DSO's per List

MS	11	19	20	24	25	24	19	14	14
OC	4	12	27	31	13	13	8	9	12
OC/N	1	1	2	6	7	1	2	3	0
Neb	0	0	0	2	7	2	1	0	0
PN	2	2	0	1	2	1	3	0	1
GC	1	1	1	1	0	2	0	1	0
Gx	29	13	4	6	8	12	36	27	19
Total	48	48	54	71	62	55	69	54	46

Chart 6

Southern Hemisphere Sky Chart

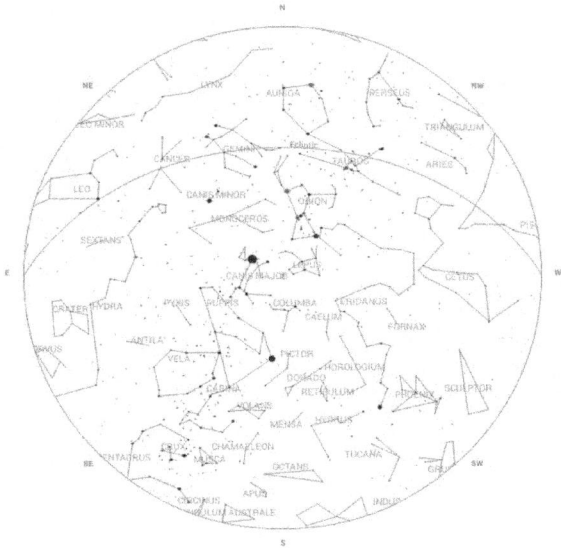

Suggested Objects for Both Hemispheres (from west to east)

Con.	Object	Type	R.A.	Dec.	List #
Per	Eta Per	MS	02h 51m	+55° 54′	3
Tau	Mel 25	OC	04h 27m	+16° 00′	4
Aur	M 38	OC	05h 29m	+35° 51′	5
Tau	M 1	SNR	05h 35m	+22° 01′	6
Ori	Del Ori	MS	05h 32m	-00° 18′	6
	Iot Ori	MS	05h 35m	-05° 55′	6
	M 78	Neb	05h 47m	+00° 05′	6
	NGC 2169	OC	06h 08m	+13° 58′	6
Mon	M 50	OC	07h 03m	-08° 23′	7
Lyn	19 Lyn	MS	07h 23m	+55° 17′	7
Gem	Alp Gem	MS	07h 35m	+31° 53′	8
Cnc	Zet Cnc	MS	08h 12m	+17° 39′	8
Pup	M 48	OC	08h 14m	-05° 45′	8

Chart 7

Northern Hemisphere Sky Chart

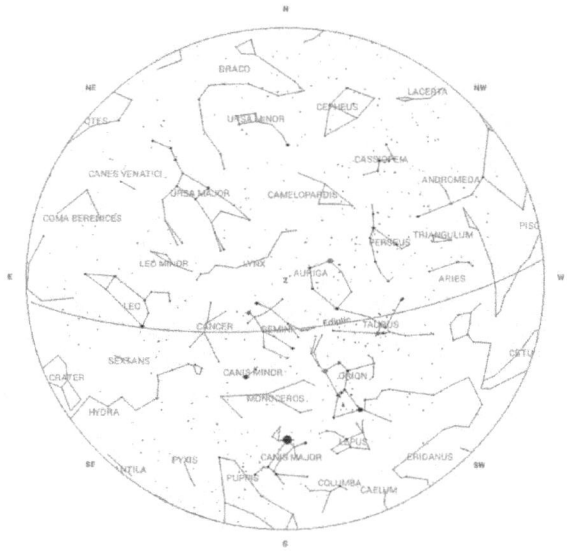

	Eastern Hemisphere			N/S		Western Hemisphere		
	Low			High	Z	High		Low
List #	11	10	9	8	7	6	5	4	3

See the individual lists for more observable objects

Number of DSO's per List

MS	18	11	19	20	24	25	24	19	14
OC	17	4	12	27	31	13	13	8	9
OC/N	3	1	1	2	6	7	1	2	3
Neb	1	0	0	0	2	7	2	1	0
PN	3	2	2	0	1	2	1	3	0
GC	0	1	1	1	1	0	2	0	1
Gx	44	29	13	4	6	8	12	36	27
Total	86	48	48	54	71	62	55	69	54

Chart 7

Southern Hemisphere Sky Chart

Suggested Objects for Both Hemispheres (from west to east)

Con.	Object	Type	R.A.	Dec.	List #
Per	Eps Per	MS	03h 58m	+40° 01'	4
Cam	1 Cam	MS	04h 32m	+53° 55'	5
Aug	M 36	OC	05h 36m	+34° 08'	6
	M 37	OC	05h 52m	+32° 33'	6
Ori	STF 855	MS	06h 09m	+02° 30'	6
Gem	20 Gem	MS	06h 32m	+17° 47'	7
	NGC 2392	PN	07h 29m	+20° 55'	7
CMa	NGC 2362	OC	07h 19m	-24° 57'	7
Pup	M 46	OC	07h 42m	-14° 49'	8
	2 Pup	MS	07h 46m	-14° 41'	8
Cnc	M 44	OC	08h 40m	+19° 40'	9
Uma	M 81	Gx S	09h 56m	+69° 04'	10
Leo	Gam Leo	MS	10h 20m	+19° 50'	10

Chart 8

Northern Hemisphere Sky Chart

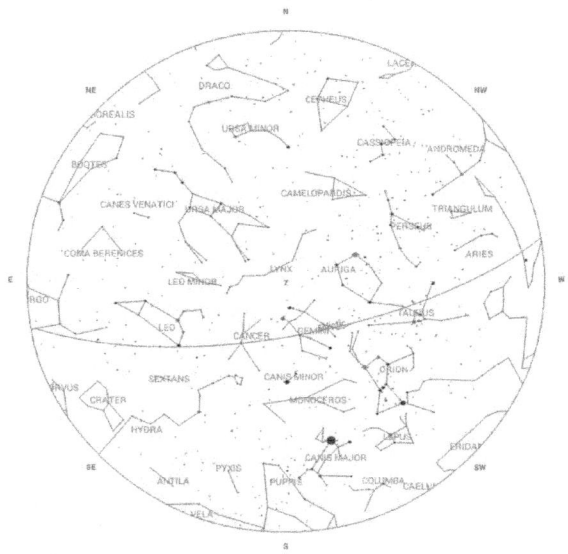

	Eastern Hemisphere			N/S	Western Hemisphere			
	Low			High	Z	High			Low
List #	12	11	10	9	8	7	6	5	4

See the individual lists for more observable objects

	Number of DSO's per List								
MS	15	18	11	19	20	24	25	24	19
OC	10	17	4	12	27	31	13	13	8
OC/N	2	3	1	1	2	6	7	1	2
Neb	1	1	0	0	0	2	7	2	1
PN	1	3	2	2	0	1	2	1	3
GC	2	0	1	1	1	1	0	2	0
Gx	86	44	29	13	4	6	8	12	36
Total	117	86	48	48	54	71	62	55	69

Chart 8

Southern Hemisphere Sky Chart

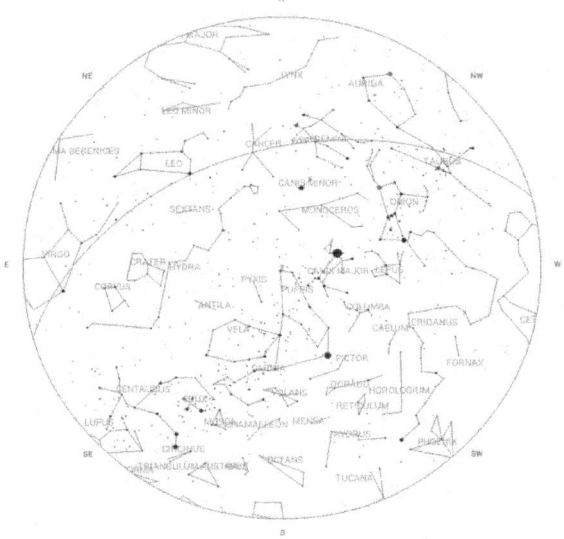

Suggested Objects for Both Hemispheres (from west to east)

Con.	Object	Type	R.A.	Dec.	List #
Ori	The Ori	MS	05h 35m	-05° 25'	6
	STF 747	MS	05h 35m	-06° 00'	6
Mon	M 50	OC	07h 03m	-08° 23'	7
Lyn	19 Lyn	MS	07h 23m	+55° 17'	7
Gem	Alp Gem	MS	07h 35m	+31° 53'	8
Cam	NGC 2403	Gx S	07h 37m	+65° 36'	8
Pup	M 93	OC	07h 44m	-23° 51'	8
Cnc	Iot Cnc	MS	08h 47m	+28° 46'	9
Hya	NGC 2342	PN	10h 25m	-18° 39'	10
	NGC 3621	Gx S	11h 18m	-32° 49'	11
Leo	M 95	Gx S	10h 44m	+11° 42'	11
	54 Leo	MS	10h 56m	+24° 45'	11
	M 66	Gx S	11h 20m	+12° 59'	11

Chart 9

Northern Hemisphere Sky Chart

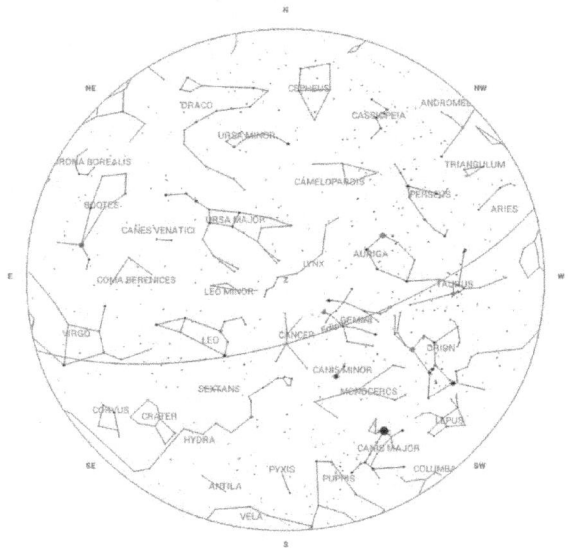

	Eastern Hemisphere				N/S	Western Hemisphere			
	Low			High	Z	High			Low
List #	13	12	11	10	9	8	7	6	5

See the individual lists for more observable objects

	Number of DSO's per List								
MS	16	15	18	11	19	20	24	25	24
OC	4	10	17	4	12	27	31	13	13
OC/N	0	2	3	1	1	2	6	7	1
Neb	0	1	1	0	0	0	2	7	2
PN	0	1	3	2	2	0	1	2	1
GC	5	2	0	1	1	1	1	0	2
Gx	84	86	44	29	13	4	6	8	12
Total	109	117	86	48	48	54	71	62	55

Chart 9

Southern Hemisphere Sky Chart

Suggested Objects for Both Hemispheres (from west to east)

Con.	Object	Type	R.A.	Dec.	List #
Ori	Lam Ori	MS	05h 35m	+09° 56'	6
	M 42	Neb	05h 35m	-05° 23'	6
Mon	Bet Mon	MS	06h 29m	-07° 02'	6
Aur	NGC 2281	OC	06h 48m	+41° 05'	7
CMa	145 CMa	MS	07h 17m	-23° 19'	7
Pup	M 46	OC	07h 42m	-14° 49'	8
	2 Pup	MS	07h 46m	-14° 41'	8
Cnc	M 67	OC	08h 51m	+11° 49'	9
UMa	M 82	Gx I	09h 56m	+69° 41'	10
Leo	M 96	Gx S	10h 47m	+11° 49'	11
	M 65	Gx S	11h 19m	+13° 06'	11
Com	M 85	Gx S	12h 25m	+18° 11'	12
Vir	M 86	Gx E	12h 26m	+12° 57'	12

Chart 10

Northern Hemisphere Sky Chart

	Eastern Hemisphere				N/S	Western Hemisphere			
	Low			High	Z	High			Low
List #	14	13	12	11	10	9	8	7	6

See the individual lists for more observable objects

	Number of DSO's per List								
MS	11	16	15	18	11	19	20	24	25
OC	5	4	10	17	4	12	27	31	13
OC/N	0	0	2	3	1	1	2	6	7
Neb	0	0	1	1	0	0	0	2	7
PN	2	0	1	3	2	2	0	1	2
GC	4	5	2	0	1	1	1	1	0
Gx	22	84	86	44	29	13	4	6	8
Total	44	109	117	86	48	48	54	71	62

Chart 10

Southern Hemisphere Sky Chart

Suggested Objects for Both Hemispheres (from west to east)

Con.	Object	Type	R.A.	Dec.	List #
CMa	M 41	OC	06h 46m	-20° 45'	7
Cnc	Zet Cnc	MS	08h 12m	+17° 39'	8
Hya	M 48	OC	08h 14m	-05° 45'	8
Cnc	Iot Cnc	MS	08h 47m	+28° 46'	9
Hya	NGC 3109	Gx S	10h 03m	-26° 10'	10
Leo	Gam Leo	MS	10h 20m	+19° 50'	10
	M 105	Gx e	10h 48m	+12° 35'	11
	NGC 3521	Gx S	11h 06m	-00° 02'	11
	NGC 3628	Gx S	11h 20m	+13° 35'	11
Vir	M 61	Gx S	12h 22m	+04° 28'	12
	M 84	Gx E	12h 25m	+12° 53'	12
Cam	32 Cam	MS	12h 49m	+83° 25'	13
Hya	M 68	GC	12h 39m	-26° 45'	13

Chart 11

Northern Hemisphere Sky Chart

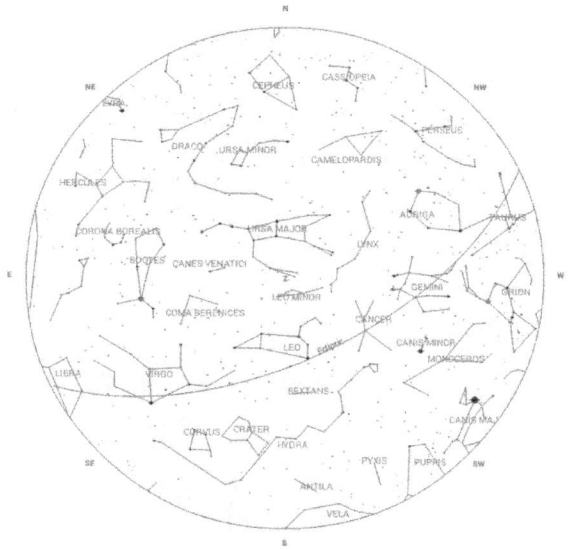

	Eastern Hemisphere			N/S	Western Hemisphere			
	Low			High	Z	High			Low
List #	15	14	13	12	11	10	9	8	7

See the individual lists for more observable objects

Number of DSO's per List

MS	18	11	16	15	18	11	19	20	24
OC	4	5	4	10	17	4	12	27	31
OC/N	0	0	0	2	3	1	1	2	6
Neb	0	0	0	1	1	0	0	0	2
PN	1	2	0	1	3	2	2	0	1
GC	5	4	5	2	0	1	1	1	1
Gx	13	22	84	86	44	29	13	4	6
Total	41	44	109	117	86	48	48	54	71

Chart 11

Southern Hemisphere Sky Chart

Suggested Objects for Both Hemispheres (from west to east)

Con.	Object	Type	R.A.	Dec.	List #
Pup	M 47	OC	07h 37m	-14° 29'	8
	Kap Pup	MS	07h 39m	-26° 48'	8
Cnc	M 44	OC	08h 40m	+19° 40'	9
UMa	M 82	Gx I	09h 56m	+69° 41'	10
Leo	M 95	Gx S	10h 44m	+11° 42'	11
	54 Leo	MS	10h 56m	+24° 45'	11
UMa	M 108	Gx S	11h 12m	+55° 40'	11
	M 97	PN	11h 15m	+55° 01'	11
Com	M 99	Gx S	12h 19m	+14° 25'	12
	M 100	Gx S	12h 23m	+15° 49'	12
CVn	M 94	Gx S	12h 51m	+41° 07'	13
Com	M 64	Gx S	12h 57m	+21° 41'	13
CVn	M 3	GC	13h 42m	+28° 23'	14

Chart 12

Northern Hemisphere Sky Chart

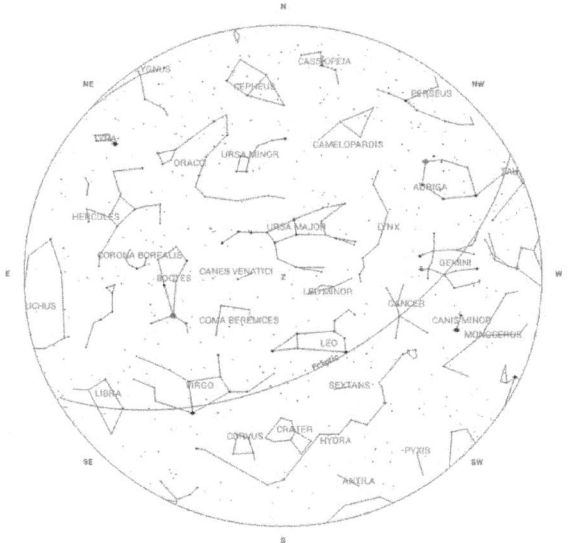

	Eastern Hemisphere			N/S	Western Hemisphere			
	Low			High	Z	High			Low
List #	16	15	14	13	12	11	10	9	8

See the individual lists for more observable objects

	Number of DSO's per List								
MS	21	18	11	16	15	18	11	19	20
OC	7	4	5	4	10	17	4	12	27
OC/N	0	0	0	0	2	3	1	1	2
Neb	0	0	0	0	1	1	0	0	0
PN	1	1	2	0	1	3	2	2	0
GC	7	5	4	5	2	0	1	1	1
Gx	4	13	22	84	86	44	29	13	4
Total	40	41	44	109	117	86	48	48	54

Chart 12

Southern Hemisphere Sky Chart

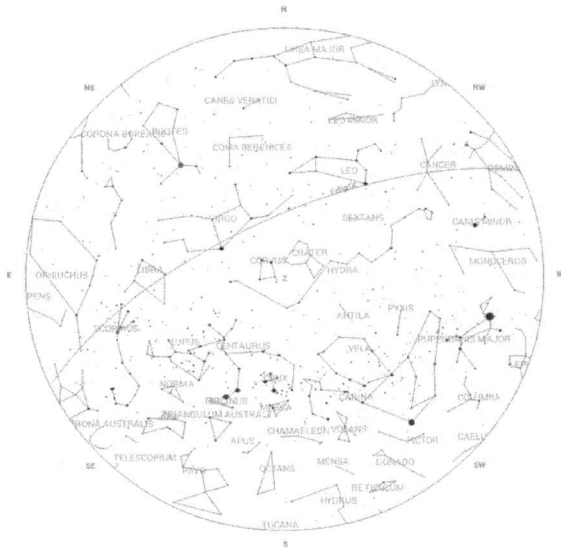

Suggested Objects for Both Hemispheres (from west to east)

Con.	Object	Type	R.A.	Dec.	List #
Hya	NGC 3242	PN	10h 25m	-18° 39'	10
Leo	M 105	Gx e	10h 48m	+12° 35'	11
	NGC 3521	Gx S	11h 06m	-00° 02'	11
	NGC 3628	Gx S	11h 20m	+13° 35'	11
UMa	M 109	Gx S	11h 58m	+53° 23'	12
Com	M 98	Gx S	12h 14m	+14° 54'	12
Dra	NGC 4236	Gx S	12h 17m	+69° 27'	12
CVn	M 106	Gx S	12h 19m	+47° 18'	12
Vir	M 49	Gx E	12h 30m	+08° 00'	12
	M 60	Gx E	12h 44m	+11° 33'	13
Com	M 63	Gx S	13h 16m	+42° 02'	13
UMa	M 101	Gx S	14h 03m	+54° 21'	14
Boo	Mu Boo	MS	15h 25m	+37° 23'	15

Chart 13

Northern Hemisphere Sky Chart

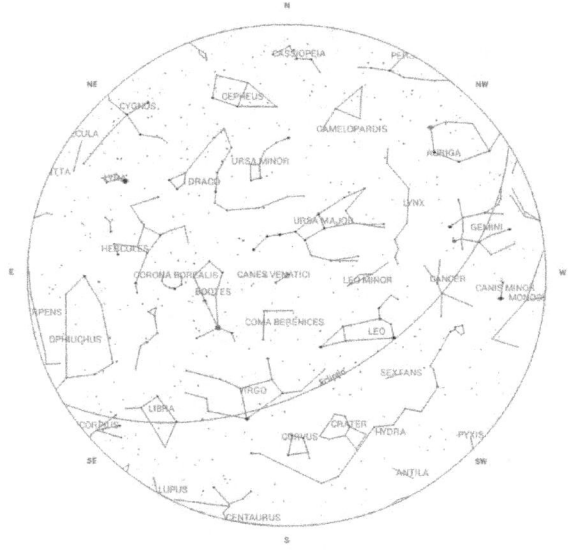

	Eastern Hemisphere				N/S	Western Hemisphere			
	Low			High	Z	High			Low
List #	17	16	15	14	13	12	11	10	9

	Number of DSO's per List								
MS	14	21	18	11	16	15	18	11	19
OC	18	7	4	5	4	10	17	4	12
OC/N	1	0	0	0	0	2	3	1	1
Neb	0	0	0	0	0	1	1	0	0
PN	1	1	1	2	0	1	3	2	2
GC	21	7	5	4	5	2	0	1	1
Gx	6	4	13	22	84	86	44	29	13
Total	61	40	41	44	109	117	86	48	48

Chart 13

Southern Hemisphere Sky Chart

Suggested Objects for Both Hemispheres (from west to east)

Con.	Object	Type	R.A.	Dec.	List #
UMa	M 81	Gx S	09h 56m	+69° 04'	10
Leo	M 96	Gx S	10h 47m	+11° 49'	11
	M 65	Gx S	11h 19m	+13° 06'	11
Com	M 99	Gx S	12h 19m	+14° 25'	12
	M 100	Gx S	12h 23m	+15° 49'	12
	M 91	Gx S	12h 35m	+14° 30'	13
Vir	M 89	Gx E	12h 36m	+12° 33'	13
	M 58	Gx S	12h 38m	+11° 49'	13
	M 59	Gx E	12h 42m	+11° 39'	13
CVn	Alp CVn	MS	12h 56m	+38° 19'	13
UMa	Zet UMa	MS	13h 24m	+54° 56'	13
Hya	M 83	Gx S	13h 37m	-29° 52'	14
Boo	Xi Boo	MS	14h 51m	+19° 06'	15

Chart 14

Northern Hemisphere Sky Chart

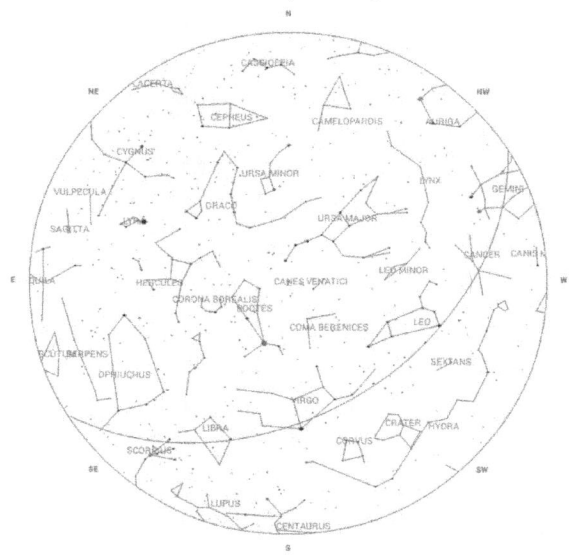

	Eastern Hemisphere				N/S	Western Hemisphere			
	Low			High	Z	High			Low
List #	18	17	16	15	14	13	12	11	10

	Number of DSO's per List								
MS	16	14	21	18	11	16	15	18	11
OC	23	18	7	4	5	4	10	17	4
OC/N	11	1	0	0	0	0	2	3	1
Neb	1	0	0	0	0	0	1	1	0
PN	4	1	1	1	2	0	1	3	2
GC	22	21	7	5	4	5	2	0	1
Gx	4	6	4	13	22	84	86	44	29
Total	81	61	40	41	44	109	117	86	48

Chart 14

Southern Hemisphere Sky Chart

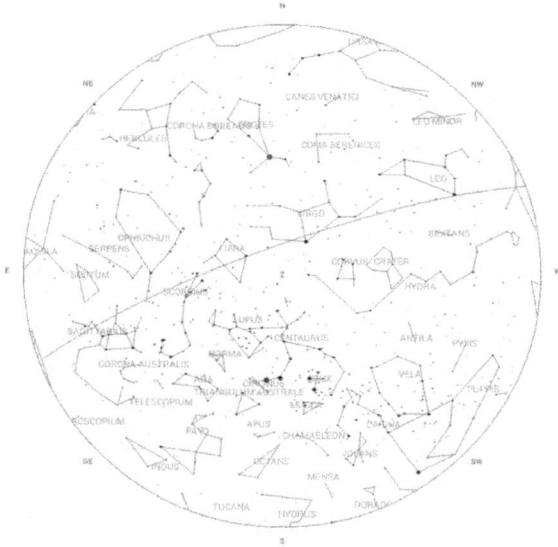

Suggested Objects for Both Hemispheres (from west to east)

Con.	Object	Type	R.A.	Dec.	List #
Hya	NGC 3621	Gx S	11h 18m	-32° 49'	11
Leo	M 66	Gx S	11h 20m	+12° 59'	11
Vir	M 61	Gx S	12h 22m	+04° 28'	12
	M 84	Gx E	12h 25m	+12° 53'	12
Com	M 88	Gx S	12h 32m	+14° 25'	13
Vir	M 90	Gx S	12h 37m	+13° 10'	13
Boo	NGC 5466	OC	14h 05m	+28° 32'	14
	Kap Boo	MS	14h 14m	+51° 47'	14
	Pi Boo	MS	14h 41m	+16° 25'	15
Ser	M 5	GC	15h 19m	+02° 05'	15
	Del Ser	MS	15h 35m	+10° 32'	16
Sco	Xi Sco	MS	16h 04m	-11° 22'	16
Dra	16 Dra	MS	16h 36m	+52° 55'	17

Chart 15

Northern Hemisphere Sky Chart

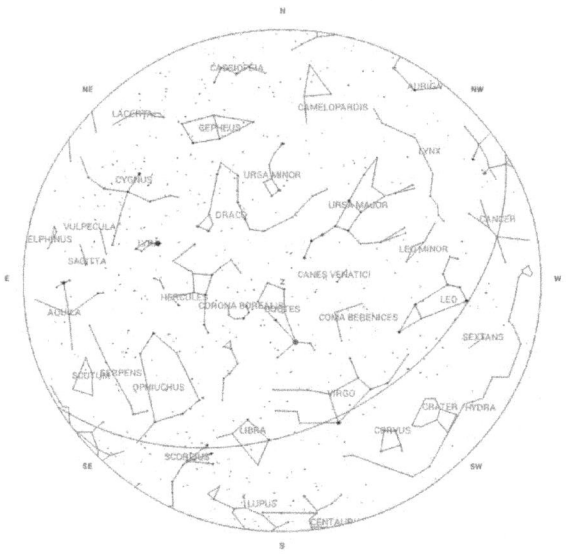

	Eastern Hemisphere			N/S	Western Hemisphere			
	Low			High	Z	High			Low
List #	19	18	17	16	15	14	13	12	11

	Number of DSO's per List								
MS	21	16	14	21	18	11	16	15	18
OC	16	23	18	7	4	5	4	10	17
OC/N	0	11	1	0	0	0	0	2	3
Neb	0	1	0	0	0	0	0	1	1
PN	1	4	1	1	1	2	0	1	3
GC	13	22	21	7	5	4	5	2	0
Gx	3	4	6	4	13	22	84	86	44
Total	54	81	61	40	41	44	109	117	86

Chart 15

Southern Hemisphere Sky Chart

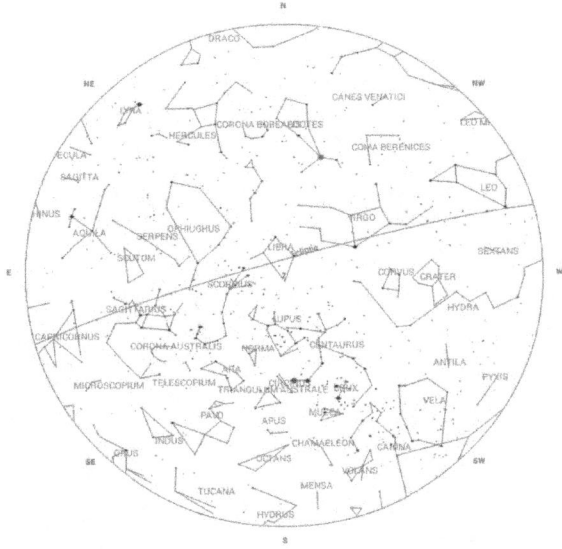

Suggested Objects for Both Hemispheres (from west to east)

Con.	Object	Type	R.A.	Dec.	List #
Com	M 85	Gx S	12h 25m	+18° 11'	12
Vir	M 86	Gx E	12h 26m	+12° 57'	12
	M 87	Gx E	12h 31m	+12° 23'	13
CVn	M 51	Gx S	13h 30m	+47° 12'	13
Hya	M 83	Gx S	13h 37m	-29° 52'	14
Dra	M 102	Gx S	15h 06m	+55° 46'	15
Lib	NGC 5897	GC	15h 17m	-21° 01'	15
CrB	Zet CrB	MS	15h 39m	+36° 38'	16
Her	Kap Her	MS	16h 08m	+17° 03'	16
	36 Her	MS	16h 41m	+04° 13'	17
Oph	M 12	GC	16h 47m	-01° 57'	17
	M 62	GC	17h 01m	-30° 07'	17
Dra	Nu Dra	MS	17h 32m	+55° 11'	18

Chart 16

Northern Hemisphere Sky Chart

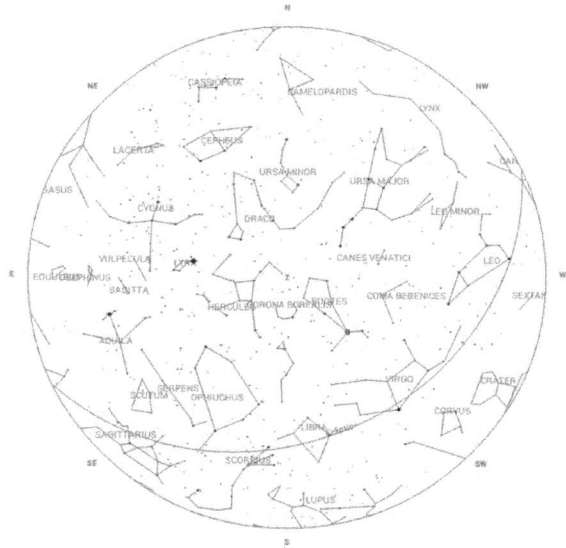

	Eastern Hemisphere				N/S	Western Hemisphere			
	Low			High	Z	High			Low
List #	20	19	18	17	16	15	14	13	12
	Number of DSO's per List								
MS	23	21	16	14	21	18	11	16	15
OC	10	16	23	18	7	4	5	4	10
OC/N	7	0	11	1	0	0	0	0	2
Neb	1	0	1	0	0	0	0	0	1
PN	7	1	4	1	1	1	2	0	1
GC	3	13	22	21	7	5	4	5	2
Gx	6	3	4	6	4	13	22	84	86
Total	57	54	81	61	40	41	44	109	117

Chart 16

Southern Hemisphere Sky Chart

Suggested Objects for Both Hemispheres (from west to east)

Con.	Object	Type	R.A.	Dec.	List #
Vir	M 104	Gx S	12h 40m	-11° 37'	13
Com	M 53	GC	13h 13m	+18° 10'	13
UMa	M 101	Gx S	14h 03m	+54° 21'	14
Boo	Pi Boo	MS	14h 41m	+16° 25'	15
Ser	M 5	GC	15h 19m	+02° 05'	15
Sco	M 80	GC	16h 17m	-22° 59'	16
	M 4	GC	16h 24m	-26° 32'	16
Oph	M 107	GC	16h 33m	-13° 03'	17
	M 10	GC	16h 57m	-04° 06'	17
Her	M 92	GC	17h 17m	+43° 08'	17
Sgr	M 23	OC	17h 57m	-18° 59'	18
Dra	40 Dra	MS	18h 00m	+80° 00'	18
Sgr	M 22	GC	18h 36m	-23° 54'	19
Sct	M 11	OC	18h 51m	-06° 16'	19

Chart 17

Northern Hemisphere Sky Chart

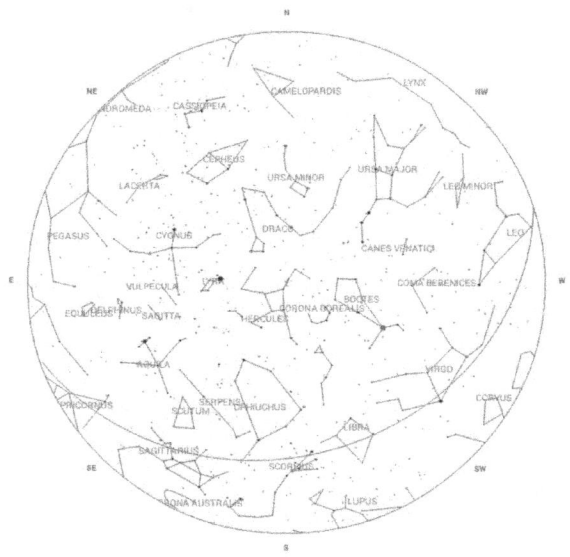

	Eastern Hemisphere				N/S	Western Hemisphere			
	Low			High	Z	High			Low
List #	21	20	19	18	17	16	15	14	13

	Number of DSO's per List								
MS	15	23	21	16	14	21	18	11	16
OC	5	10	16	23	18	7	4	5	4
OC/N	1	7	0	11	1	0	0	0	0
Neb	2	1	0	1	0	0	0	0	0
PN	2	7	1	4	1	1	1	2	0
GC	4	3	13	22	21	7	5	4	5
Gx	5	6	3	4	6	4	13	22	84
Total	34	57	54	81	61	40	41	44	109

Chart 17

Southern Hemisphere Sky Chart

Suggested Objects for Both Hemispheres (from west to east)

Con.	Object	Type	R.A.	Dec.	List #
CVn	M 3	GC	13h 42m	+28° 23'	14
Boo	Xi Boo	MS	14h 51m	+19° 06'	15
CrB	Zet CrB	MS	15h 39m	+36° 38'	16
Her	Kap Her	MS	16h 08m	+17° 03'	16
Oph	M 19	GC	17h 03m	-26° 16'	17
	36 Oph	MS	17h 15m	-26° 36'	17
	Omi Oph	MS	17h 18m	-24° 17'	17
	M 9	GC	17h 19m	-18° 31'	17
Sgr	M 21	OC	18h 04m	-22° 29'	18
Ser	M 16	OC/N	18h 19m	-13° 49'	18
Sgr	M 25	OC	18h 32m	-19° 15'	19
	NGC 6723	GC	19h 00m	-36° 38'	19
Aql	57 Aql	MS	19h 55m	-08° 14'	20

Chart 18

Northern Hemisphere Sky Chart

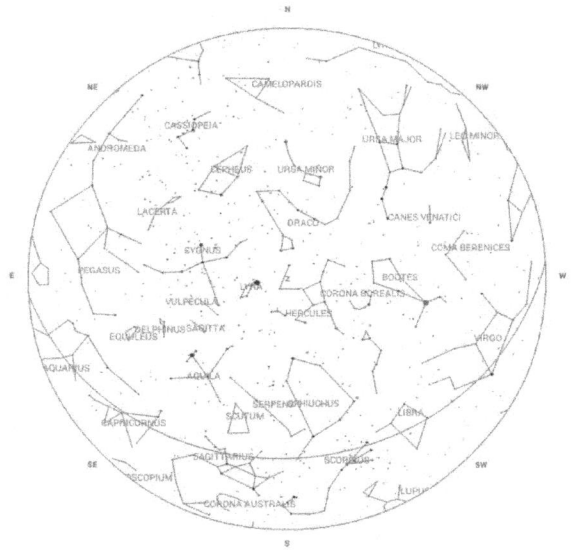

	Eastern Hemisphere			N/S	Western Hemisphere			
	Low			High	Z	High			Low
List #	22	21	20	19	18	17	16	15	14

	Number of DSO's per List								
MS	15	15	23	21	16	14	21	18	11
OC	7	5	10	16	23	18	7	4	5
OC/N	3	1	7	0	11	1	0	0	0
Neb	0	2	1	0	1	0	0	0	0
PN	2	2	7	1	4	1	1	1	2
GC	2	4	3	13	22	21	7	5	4
Gx	10	5	6	3	4	6	4	13	22
Total	39	34	57	54	81	61	40	41	44

Chart 18

Southern Hemisphere Sky Chart

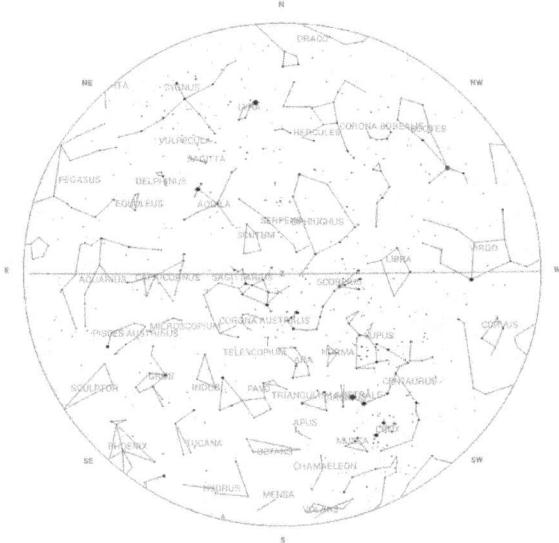

Suggested Objects for Both Hemispheres (from west to east)

Con.	Object	Type	R.A.	Dec.	List #
Boo	Mu Boo	MS	15h 25m	+37° 23'	15
Ser	Del Ser	MS	15h 35m	+10° 32'	16
Sco	Xi Sco	MS	16h 04m	-11° 22'	16
Oph	M 107	GC	16h 33m	-13° 03'	17
	M 10	GC	16h 57m	-04° 06'	17
Her	M 92	GC	17h 17m	+43° 08'	17
Sgr	M 20	Neb	18h 02m	-23° 02'	18
	M 24	OC	18h 18m	-18° 24'	18
	M 18	OC/N	18h 20m	-17° 06'	18
	M 28	GC	18h 25m	-24° 52'	18
Oph	M 14	GC	17h 38m	-03° 15'	18
Sgr	M 69	GC	18h 31m	-32° 21'	19
	M 54	GC	18h 55m	-30° 29'	19
Cyg	16 Cyg	MS	19h 42m	+50° 32'	20

Chart 19

Northern Hemisphere Sky Chart

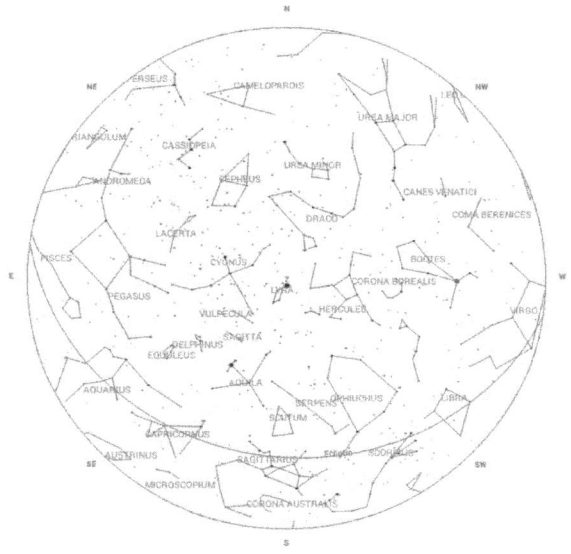

	Eastern Hemisphere			N/S	Western Hemisphere			
	Low			High	High			Low
List #	23	22	21	20	19	18	17	16	15

	Number of DSO's per List							
MS	13	15	15	23	21	16	14	21	18
OC	2	7	5	10	16	23	18	7	4
OC/N	3	3	1	7	0	11	1	0	0
Neb	0	0	2	1	0	1	0	0	0
PN	1	2	2	7	1	4	1	1	1
GC	0	2	4	3	13	22	21	7	5
Gx	19	10	5	6	3	4	6	4	13

Total	38	39	34	57	54	81	61	40	41

Chart 19

Southern Hemisphere Sky Chart

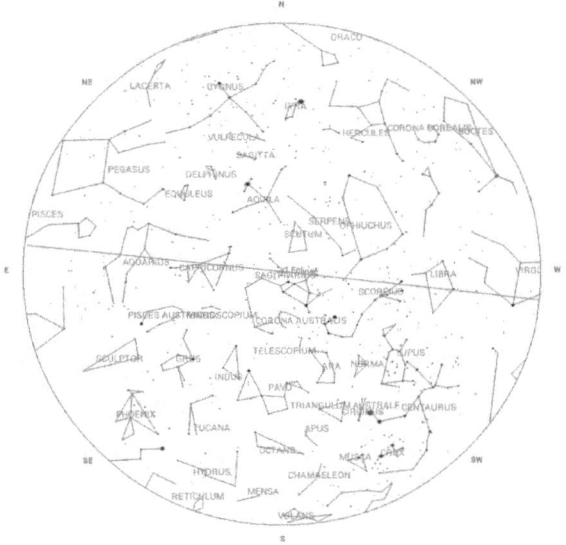

Suggested Objects for Both Hemispheres (from west to east)

Con.	Object	Type	R.A.	Dec.	List #
Sco	Nu Sco	MS	16h 12m	-19° 28'	16
Her	36 Her	MS	16h 41m	+04° 13'	17
Oph	M 12	GC	16h 47m	-01° 57'	17
	M 62	GC	17h 01m	-30° 07'	17
	61 Oph	MS	17h 45m	+02° 35'	18
Sgr	M 17	OC/N	18h 21m	-16° 10'	18
	M 70	GC	18h 43m	-32° 18'	19
Sct	M 11	OC	18h 51m	-06° 16'	11
Lyr	M 57	PN	18h 54m	+33° 02'	19
	M 56	GC	19h 17m	+30°11'	19
Sge	M 71	GC	19h 54m	+18° 47'	20
Vul	M 27	PN	20h 00m	+22° 43'	20
Cyg	61 Cyg	MS	21h 07m	+38° 45'	21
	M 39	OC	21h 32m	+48° 27'	22

Chart 20

Northern Hemisphere Sky Chart

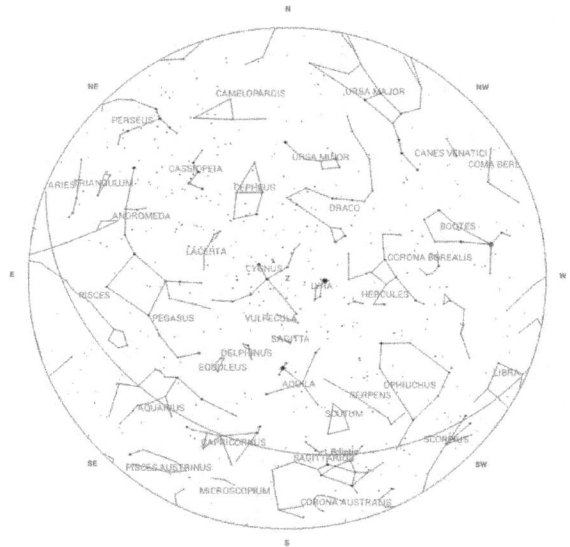

	Eastern Hemisphere				N/S	Western Hemisphere			
	Low			High	Z	High			Low
List #	24	23	22	21	20	19	18	17	16

	Number of DSO's per List								
MS	6	13	15	15	23	21	16	14	21
OC	0	2	7	5	10	16	23	18	7
OC/N	0	3	3	1	7	0	11	1	0
Neb	0	0	0	2	1	0	1	0	0
PN	0	1	2	2	7	1	4	1	1
GC	1	0	2	4	3	13	22	21	7
Gx	11	19	10	5	6	3	4	6	4
Total	18	38	39	34	57	54	81	61	40

Chart 20

Southern Hemisphere Sky Chart

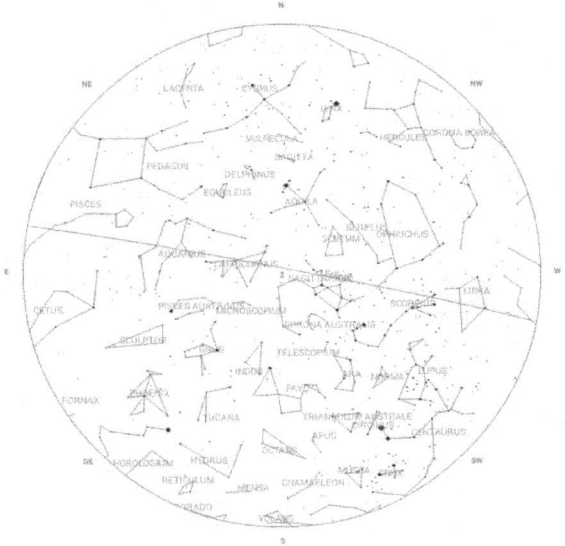

Suggested Objects for Both Hemispheres (from west to east)

Con.	Object	Type	R.A.	Dec.	List #
Dra	16 Dra	MS	16h 36m	+52° 55'	17
Her	M 13	GC	16h 42m	+36° 28'	17
Sgr	M 8	OC/N	18h 03m	-24° 23'	18
	M 69	GC	18h 31m	-32° 21'	19
	M 54	GC	18h 55m	-30° 29'	19
	NGC 6822	Gx I	19h 45m	-14° 49'	20
	M 75	GC	20h 06m	-21° 55'	20
Cyg	Bet Cyg	MS	19h 31m	+27° 58'	20
	NGC 6826	PN	19h 45m	+50° 32'	21
	NGC 7027	PN	21h 07m	+42° 14'	21
Aqr	NGC 7009	PN	21h 04m	-11° 22'	21
Aqr	M 2	GC	21h 33m	-00° 49'	22
Lac	8 Lac	MS	22h 36m	+39° 38'	23

Chart 21

Northern Hemisphere Sky Chart

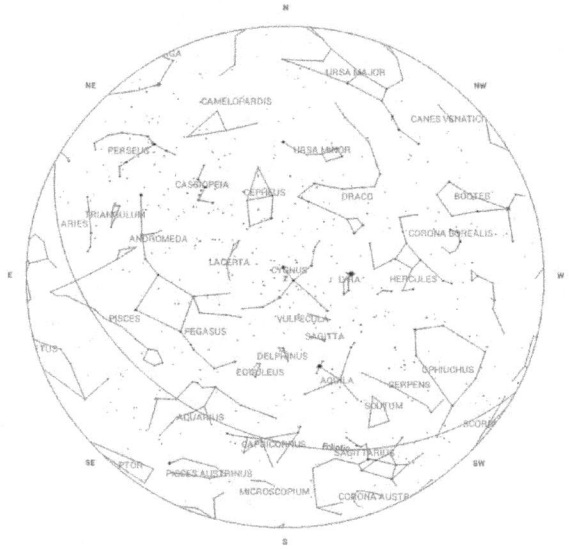

	Eastern Hemisphere				N/S Z	Western Hemisphere			
	Low			High		High			Low
List #	1	24	23	22	21	20	19	18	17

	Number of DSO's per List								
MS	15	6	13	15	15	23	21	16	14
OC	6	0	2	7	5	10	16	23	18
OC/N	1	0	3	3	1	7	0	11	1
Neb	1	0	0	0	2	1	0	1	0
PN	0	0	1	2	2	7	1	4	1
GC	2	1	0	2	4	3	13	22	21
Gx	21	11	19	10	5	6	3	4	6
Total	46	18	38	39	34	57	54	81	61

Chart 21

Southern Hemisphere Sky Chart

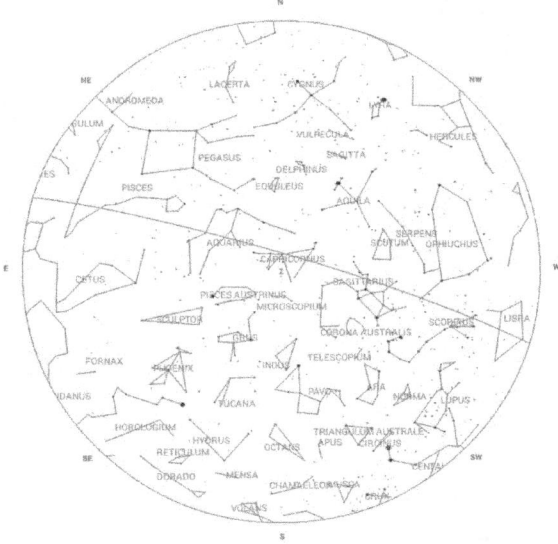

Suggested Objects for Both Hemispheres (from west to east)

Con.	Object	Type	R.A.	Dec.	List #
Dra	Psi¹ Dra	MS	17h 42m	+72° 09'	18
Oph	70 Oph	MS	18h 06m	+02° 30'	18
	NGC 6633	OC	18h 27m	+06° 31'	18
Sgr	NGC 6530	OC/N	18h 05m	-24° 21'	18
	M 25	OC	18h 32m	-19° 15'	19
Sge	M 71	GC	19h 54m	+18° 47'	20
Vul	M 27	PN	20h 00m	+22° 43'	20
Cyg	NGC 6946	Gx S	20h 35m	+60° 09'	21
	NGC 6992	SNR	20h 56m	+31° 45'	21
	NGC 7000	Neb	21h 02m	+44° 12'	21
Aqr	M 72	GC	20h 53m	-12° 32'	21
Cap	M 30	GC	21h 40m	-23° 11'	22
Cas	M 52	OC	23h 25m	+61° 36'	23

Chart 22

Northern Hemisphere Sky Chart

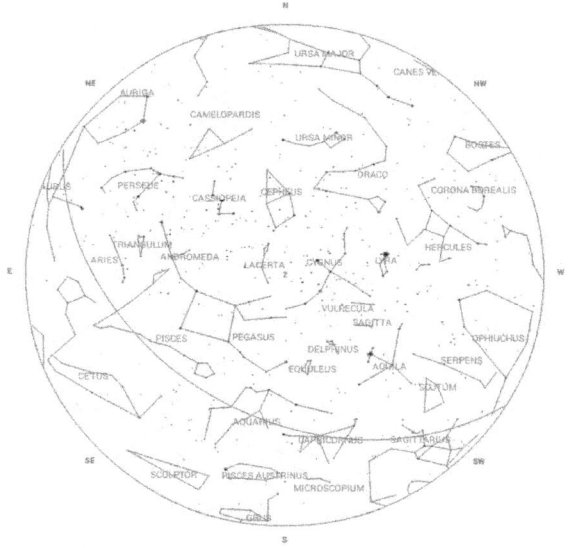

	Eastern Hemisphere				N/S Z	Western Hemisphere			
	Low			High		High			Low
List #	2	1	24	23	22	21	20	19	18

	Number of DSO's per List								
MS	14	15	6	13	15	15	23	21	16
OC	12	6	0	2	7	5	10	16	23
OC/N	0	1	0	3	3	1	7	0	11
Neb	0	1	0	0	0	2	1	0	1
PN	1	0	0	1	2	2	7	1	4
GC	0	2	1	0	2	4	3	13	22
Gx	19	21	11	19	10	5	6	3	4
Total	46	46	18	38	39	34	57	54	81

Chart 22

Southern Hemisphere Sky Chart

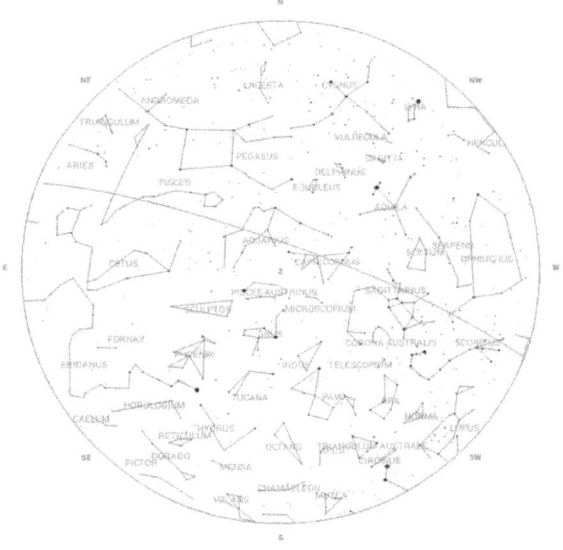

Suggested Objects for Both Hemispheres (from west to east)

Con.	Object	Type	R.A.	Dec.	List #
Lyr	Eps Lyr	MS	18h 44m	+39° 40′	19
Sct	M 11	OC	18h 51m	-06° 16′	19
Sgr	M 55	GC	19h 40m	-30° 58′	20
Cyg	16 Cyg	MS	19h 42m	+50° 32′	20
	NGC 7027	PN	21h 07m	+42° 14′	21
Aqr	NGC 7009	PN	21h 04m	-11° 22′	21
	NGC 7293	PN	22h 30m	-20° 50′	22
Cep	STF 2840	MS	21h 52m	+55° 48′	22
And	NGC 7662	PN	23h 26m	+42° 32′	23
	M 31	Gx S	00h 43m	+41° 16′	1
Cas	Eta Cas	MS	00h 49m	+57° 49′	1
Psc	Psi¹ Psc	MS	01h 06m	+21° 28′	1
	Zet Psc	MS	01h 14m	+07° 35′	1

Chart 23

Northern Hemisphere Sky Chart

	Eastern Hemisphere				N/S	Western Hemisphere			
	Low			High	Z	High			Low
List #	3	2	1	24	23	22	21	20	19

Number of DSO's per List

MS	14	14	15	6	13	15	15	23	21
OC	9	12	6	0	2	7	5	10	16
OC/N	3	0	1	0	3	3	1	7	0
Neb	0	0	1	0	0	0	2	1	0
PN	0	1	0	0	1	2	2	7	1
GC	1	0	2	1	0	2	4	3	13
Gx	27	19	21	11	19	10	5	6	3
Total	54	46	46	18	38	39	34	57	54

Chart 23

Southern Hemisphere Sky Chart

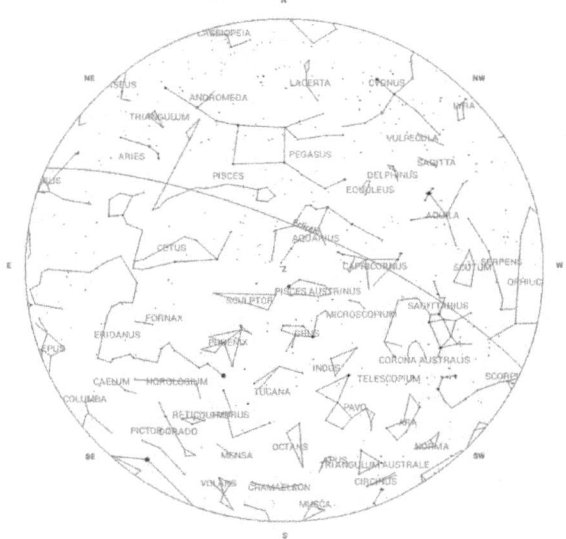

Suggested Objects for Both Hemispheres (from west to east)

Con.	Object	Type	R.A.	Dec.	List #
Aql	57 Aql	MS	19h 55m	-08° 14'	20
Cap	Omi Cap	MS	20h 30m	-18° 35'	20
Cyg	M 29	OC	20h 24m	+38° 30'	20
Peg	M 15	GC	21h 30m	+12° 10'	21
Cap	M 30	GC	21h 40m	-23° 11'	22
Peg	NGC 7331	Gx S	22h 37m	+34° 25'	23
Cet	NGC 247	Gx S	00h 48m	-20° 46'	1
Cas	NGC 225	OC	00h 44m	+61° 47'	1
	NGC 457	OC	01h 20m	+58° 17'	1
Ari	Gam Ari	MS	01h 54m	+19° 18'	2
	Lam Ari	MS	01h 58m	+23° 36'	2
And	Gam And	MS	02h 04m	+42° 20'	2
Per	NGC 869/884	OC	02h 19m	+57° 08'	2

Chart 24

Northern Hemisphere Sky Chart

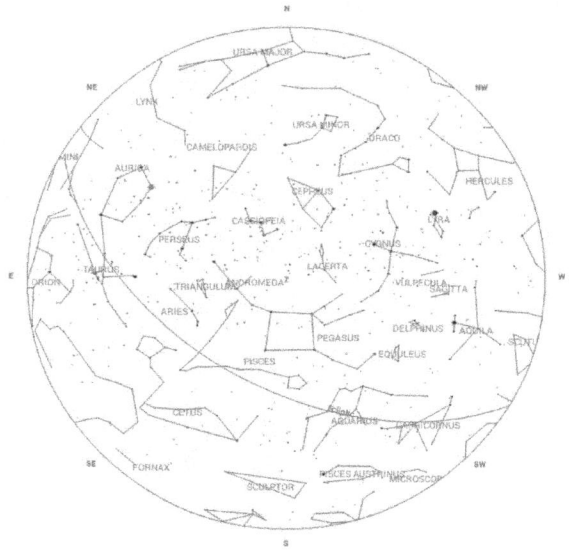

	Eastern Hemisphere				N/S	Western Hemisphere			
	Low			High	Z	High			Low
List #	4	3	2	1	24	23	22	21	20

	Number of DSO's per List								
MS	19	14	14	15	6	13	15	15	23
OC	8	9	12	6	0	2	7	5	10
OC/N	2	3	0	1	0	3	3	1	7
Neb	1	0	0	1	0	0	0	2	1
PN	3	0	1	0	0	1	2	2	7
GC	0	1	0	2	1	0	2	4	3
Gx	36	27	19	21	11	19	10	5	6
Total	69	54	46	46	18	38	39	34	57

Chart 24

Southern Hemisphere Sky Chart

Suggested Objects for Both Hemispheres (from west to east)

Con.	Object	Type	R.A.	Dec.	List #
Cyg	61 Cyg	MS	21h 07m	+38° 45'	21
Aqr	M 2	GC	21h 33m	-00° 49'	22
And	NGC 7662	PN	23h 26m	+42° 32'	23
Scl	NGC 55	Gx S	00h 15m	-39° 13'	24
	NGC 253	Gx S	00h 48m	-25°17'	1
	NGC 288	GC	00h 52m	-26° 35'	1
Psc	M 74	Gx S	01h 37m	+15° 47'	2
Tri	M 33	Gx S	01h 34m	+30° 40'	2
	NGC 925	Gx S	02h 27m	+33° 35'	2
And	NGC 752	OC	01h 58m	+37° 51'	2
Per	M 76	PN	01h 42m	+51° 34'	2
	Eta Per	MS	02h 51m	+55° 54'	3

The Detailed Lists

Early January

	Eastern Hemisphere				Zenith	Western Hemisphere			
	Low			High		High			Low
6p	5	4	3	2	1	24	23	22	21
7p	6	5	4	3	2	1	24	23	22
8p	7	6	5	4	3	2	1	24	23
9p	8	7	6	5	4	3	2	1	24
10p	9	8	7	6	5	4	3	2	1
11p	10	9	8	7	6	5	4	3	2
12a	11	10	9	8	7	6	5	4	3
1a	12	11	10	9	8	7	6	5	4
2a	13	12	11	10	9	8	7	6	5
3a	14	13	12	11	10	9	8	7	6
4a	15	14	13	12	11	10	9	8	7
5a	16	15	14	13	12	11	10	9	8
6a	17	16	15	14	13	12	11	10	9

Late January

	Eastern Hemisphere				Zenith	Western Hemisphere			
	Low			High		High			Low
6p	6	5	4	3	2	1	24	23	22
7p	7	6	5	4	3	2	1	24	23
8p	8	7	6	5	4	3	2	1	24
9p	9	8	7	6	5	4	3	2	1
10p	10	9	8	7	6	5	4	3	2
11p	11	10	9	8	7	6	5	4	3
12a	12	11	10	9	8	7	6	5	4
1a	13	12	11	10	9	8	7	6	5
2a	14	13	12	11	10	9	8	7	6
3a	15	14	13	12	11	10	9	8	7
4a	16	15	14	13	12	11	10	9	8
5a	17	16	15	14	13	12	11	10	9
6a	18	17	16	15	14	13	12	11	10

Early February

	Eastern Hemisphere				Zenith	Western Hemisphere			
	Low			High	High				Low
6p	7	6	5	4	3	2	1	24	23
7p	8	7	6	5	4	3	2	1	24
8p	9	8	7	6	5	4	3	2	1
9p	10	9	8	7	6	5	4	3	2
10p	11	10	9	8	7	6	5	4	3
11p	12	11	10	9	8	7	6	5	4
12a	13	12	11	10	9	8	7	6	5
1a	14	13	12	11	10	9	8	7	6
2a	15	14	13	12	11	10	9	8	7
3a	16	15	14	13	12	11	10	9	8
4a	17	16	15	14	13	12	11	10	9
5a	18	17	16	15	14	13	12	11	10
6a	19	18	17	16	15	14	13	12	11

Late February

	Eastern Hemisphere				Zenith	Western Hemisphere			
	Low			High	High				Low
6p	8	7	6	5	4	3	2	1	24
7p	9	8	7	6	5	4	3	2	1
8p	10	9	8	7	6	5	4	3	2
9p	11	10	9	8	7	6	5	4	3
10p	12	11	10	9	8	7	6	5	4
11p	13	12	11	10	9	8	7	6	5
12a	14	13	12	11	10	9	8	7	6
1a	15	14	13	12	11	10	9	8	7
2a	16	15	14	13	12	11	10	9	8
3a	17	16	15	14	13	12	11	10	9
4a	18	17	16	15	14	13	12	11	10
5a	19	18	17	16	15	14	13	12	11
6a	20	19	18	17	16	15	14	13	12

Early March

	Eastern Hemisphere				Zenith	Western Hemisphere			
	Low			High		High			Low
6p	9	8	7	6	5	4	3	2	1
7p	10	9	8	7	6	5	4	3	2
8p	11	10	9	8	7	6	5	4	3
9p	12	11	10	9	8	7	6	5	4
10p	13	12	11	10	9	8	7	6	5
11p	14	13	12	11	10	9	8	7	6
12a	15	14	13	12	11	10	9	8	7
1a	16	15	14	13	12	11	10	9	8
2a	17	16	15	14	13	12	11	10	9
3a	18	17	16	15	14	13	12	11	10
4a	19	18	17	16	15	14	13	12	11
5a	20	19	18	17	16	15	14	13	12
6a	21	20	19	18	17	16	15	14	13

Late March

	Eastern Hemisphere				Zenith	Western Hemisphere			
	Low			High		High			Low
6p	10	9	8	7	6	5	4	3	2
7p	11	10	9	8	7	6	5	4	3
8p	12	11	10	9	8	7	6	5	4
9p	13	12	11	10	9	8	7	6	5
10p	14	13	12	11	10	9	8	7	6
11p	15	14	13	12	11	10	9	8	7
12a	16	15	14	13	12	11	10	9	8
1a	17	16	15	14	13	12	11	10	9
2a	18	17	16	15	14	13	12	11	10
3a	19	18	17	16	15	14	13	12	11
4a	20	19	18	17	16	15	14	13	12
5a	21	20	19	18	17	16	15	14	13
6a	22	21	20	19	18	17	16	15	14

Early April

	Eastern Hemisphere				Zenith	Western Hemisphere			
	Low			High		High			Low
6p	11	10	9	8	7	6	5	4	3
7p	12	11	10	9	8	7	6	5	4
8p	13	12	11	10	9	8	7	6	5
9p	14	13	12	11	10	9	8	7	6
10p	15	14	13	12	11	10	9	8	7
11p	16	15	14	13	12	11	10	9	8
12a	17	16	15	14	13	12	11	10	9
1a	18	17	16	15	14	13	12	11	10
2a	19	18	17	16	15	14	13	12	11
3a	20	19	18	17	16	15	14	13	12
4a	21	20	19	18	17	16	15	14	13
5a	22	21	20	19	18	17	16	15	14
6a	23	22	21	20	19	18	17	16	15

Late April

	Eastern Hemisphere				Zenith	Western Hemisphere			
	Low			High		High			Low
6p	12	11	10	9	8	7	6	5	4
7p	13	12	11	10	9	8	7	6	5
8p	14	13	12	11	10	9	8	7	6
9p	15	14	13	12	11	10	9	8	7
10p	16	15	14	13	12	11	10	9	8
11p	17	16	15	14	13	12	11	10	9
12a	18	17	16	15	14	13	12	11	10
1a	19	18	17	16	15	14	13	12	11
2a	20	19	18	17	16	15	14	13	12
3a	21	20	19	18	17	16	15	14	13
4a	22	21	20	19	18	17	16	15	14
5a	23	22	21	20	19	18	17	16	15
6a	24	23	22	21	20	19	18	17	16

Early May

	Eastern Hemisphere				Zenith	Western Hemisphere			
	Low			High		High			Low
6p	13	12	11	10	9	8	7	6	5
7p	14	13	12	11	10	9	8	7	6
8p	15	14	13	12	11	10	9	8	7
9p	16	15	14	13	12	11	10	9	8
10p	17	16	15	14	13	12	11	10	9
11p	18	17	16	15	14	13	12	11	10
12a	19	18	17	16	15	14	13	12	11
1a	20	19	18	17	16	15	14	13	12
2a	21	20	19	18	17	16	15	14	13
3a	22	21	20	19	18	17	16	15	14
4a	23	22	21	20	19	18	17	16	15
5a	24	23	22	21	20	19	18	17	16
6a	1	24	23	22	21	20	19	18	17

Late May

	Eastern Hemisphere				Zenith	Western Hemisphere			
	Low			High		High			Low
6p	14	13	12	11	10	9	8	7	6
7p	15	14	13	12	11	10	9	8	7
8p	16	15	14	13	12	11	10	9	8
9p	17	16	15	14	13	12	11	10	9
10p	18	17	16	15	14	13	12	11	10
11p	19	18	17	16	15	14	13	12	11
12a	20	19	18	17	16	15	14	13	12
1a	21	20	19	18	17	16	15	14	13
2a	22	21	20	19	18	17	16	15	14
3a	23	22	21	20	19	18	17	16	15
4a	24	23	22	21	20	19	18	17	16
5a	1	24	23	22	21	20	19	18	17
6a	2	1	24	23	22	21	20	19	18

Early June

	Eastern Hemisphere				Zenith	Western Hemisphere			
	Low			High	High				Low
6p	15	14	13	12	11	10	9	8	7
7p	16	15	14	13	12	11	10	9	8
8p	17	16	15	14	13	12	11	10	9
9p	18	17	16	15	14	13	12	11	10
10p	19	18	17	16	15	14	13	12	11
11p	20	19	18	17	16	15	14	13	12
12a	21	20	19	18	17	16	15	14	13
1a	22	21	20	19	18	17	16	15	14
2a	23	22	21	20	19	18	17	16	15
3a	24	23	22	21	20	19	18	17	16
4a	1	24	23	22	21	20	19	18	17
5a	2	1	24	23	22	21	20	19	18
6a	3	2	1	24	23	22	21	20	19

Late June

	Eastern Hemisphere				Zenith	Western Hemisphere			
	Low			High	High				Low
6p	16	15	14	13	12	11	10	9	8
7p	17	16	15	14	13	12	11	10	9
8p	18	17	16	15	14	13	12	11	10
9p	19	18	17	16	15	14	13	12	11
10p	20	19	18	17	16	15	14	13	12
11p	21	20	19	18	17	16	15	14	13
12a	22	21	20	19	18	17	16	15	14
1a	23	22	21	20	19	18	17	16	15
2a	24	23	22	21	20	19	18	17	16
3a	1	24	23	22	21	20	19	18	17
4a	2	1	24	23	22	21	20	19	18
5a	3	2	1	24	23	22	21	20	19
6a	4	3	2	1	24	23	22	21	20

Early July

	Eastern Hemisphere				Zenith	Western Hemisphere			
	Low			High		High			Low
6p	17	16	15	14	13	12	11	10	9
7p	18	17	16	15	14	13	12	11	10
8p	19	18	17	16	15	14	13	12	11
9p	20	19	18	17	16	15	14	13	12
10p	21	20	19	18	17	16	15	14	13
11p	22	21	20	19	18	17	16	15	14
12a	23	22	21	20	19	18	17	16	15
1a	24	23	22	21	20	19	18	17	16
2a	1	24	23	22	21	20	19	18	17
3a	2	1	24	23	22	21	20	19	18
4a	3	2	1	24	23	22	21	20	19
5a	4	3	2	1	24	23	22	21	20
6a	5	4	3	2	1	24	23	22	21

Late July

	Eastern Hemisphere				Zenith	Western Hemisphere			
	Low			High		High			Low
6p	18	17	16	15	14	13	12	11	10
7p	19	18	17	16	15	14	13	12	11
8p	20	19	18	17	16	15	14	13	12
9p	21	20	19	18	17	16	15	14	13
10p	22	21	20	19	18	17	16	15	14
11p	23	22	21	20	19	18	17	16	15
12a	24	23	22	21	20	19	18	17	16
1a	1	24	23	22	21	20	19	18	17
2a	2	1	24	23	22	21	20	19	18
3a	3	2	1	24	23	22	21	20	19
4a	4	3	2	1	24	23	22	21	20
5a	5	4	3	2	1	24	23	22	21
6a	6	5	4	3	2	1	24	23	22

Early August

	Eastern Hemisphere				Zenith	Western Hemisphere			
	Low			High		High			Low
6p	19	18	17	16	15	14	13	12	11
7p	20	19	18	17	16	15	14	13	12
8p	21	20	19	18	17	16	15	14	13
9p	22	21	20	19	18	17	16	15	14
10p	23	22	21	20	19	18	17	16	15
11p	24	23	22	21	20	19	18	17	16
12a	1	24	23	22	21	20	19	18	17
1a	2	1	24	23	22	21	20	19	18
2a	3	2	1	24	23	22	21	20	19
3a	4	3	2	1	24	23	22	21	20
4a	5	4	3	2	1	24	23	22	21
5a	6	5	4	3	2	1	24	23	22
6a	7	6	5	4	3	2	1	24	23

Late August

	Eastern Hemisphere				Zenith	Western Hemisphere			
	Low			High		High			Low
6p	20	19	18	17	16	15	14	13	12
7p	21	20	19	18	17	16	15	14	13
8p	22	21	20	19	18	17	16	15	14
9p	23	22	21	20	19	18	17	16	15
10p	24	23	22	21	20	19	18	17	16
11p	1	24	23	22	21	20	19	18	17
12a	2	1	24	23	22	21	20	19	18
1a	3	2	1	24	23	22	21	20	19
2a	4	3	2	1	24	23	22	21	20
3a	5	4	3	2	1	24	23	22	21
4a	6	5	4	3	2	1	24	23	22
5a	7	6	5	4	3	2	1	24	23
6a	8	7	6	5	4	3	2	1	24

Early September

	Eastern Hemisphere				Zenith	Western Hemisphere			
	Low			High		High			Low
6p	21	20	19	18	17	16	15	14	13
7p	22	21	20	19	18	17	16	15	14
8p	23	22	21	20	19	18	17	16	15
9p	24	23	22	21	20	19	18	17	16
10p	1	24	23	22	21	20	19	18	17
11p	2	1	24	23	22	21	20	19	18
12a	3	2	1	24	23	22	21	20	19
1a	4	3	2	1	24	23	22	21	20
2a	5	4	3	2	1	24	23	22	21
3a	6	5	4	3	2	1	24	23	22
4a	7	6	5	4	3	2	1	24	23
5a	8	7	6	5	4	3	2	1	24
6a	9	8	7	6	5	4	3	2	1

Late September

	Eastern Hemisphere				Zenith	Western Hemisphere			
	Low			High		High			Low
6p	22	21	20	19	18	17	16	15	14
7p	23	22	21	20	19	18	17	16	15
8p	24	23	22	21	20	19	18	17	16
9p	1	24	23	22	21	20	19	18	17
10p	2	1	24	23	22	21	20	19	18
11p	3	2	1	24	23	22	21	20	19
12a	4	3	2	1	24	23	22	21	20
1a	5	4	3	2	1	24	23	22	21
2a	6	5	4	3	2	1	24	23	22
3a	7	6	5	4	3	2	1	24	23
4a	8	7	6	5	4	3	2	1	24
5a	9	8	7	6	5	4	3	2	1
6a	10	9	8	7	6	5	4	3	2

Early October

	Eastern Hemisphere				Zenith	Western Hemisphere			
	Low			High	High				Low
6p	23	22	21	20	19	18	17	16	15
7p	24	23	22	21	20	19	18	17	16
8p	1	24	23	22	21	20	19	18	17
9p	2	1	24	23	22	21	20	19	18
10p	3	2	1	24	23	22	21	20	19
11p	4	3	2	1	24	23	22	21	20
12a	5	4	3	2	1	24	23	22	21
1a	6	5	4	3	2	1	24	23	22
2a	7	6	5	4	3	2	1	24	23
3a	8	7	6	5	4	3	2	1	24
4a	9	8	7	6	5	4	3	2	1
5a	10	9	8	7	6	5	4	3	2
6a	11	10	9	8	7	6	5	4	3

Late October

	Eastern Hemisphere				Zenith	Western Hemisphere			
	Low			High	High				Low
6p	24	23	22	21	20	19	18	17	16
7p	1	24	23	22	21	20	19	18	17
8p	2	1	24	23	22	21	20	19	18
9p	3	2	1	24	23	22	21	20	19
10p	4	3	2	1	24	23	22	21	20
11p	5	4	3	2	1	24	23	22	21
12a	6	5	4	3	2	1	24	23	22
1a	7	6	5	4	3	2	1	24	23
2a	8	7	6	5	4	3	2	1	24
3a	9	8	7	6	5	4	3	2	1
4a	10	9	8	7	6	5	4	3	2
5a	11	10	9	8	7	6	5	4	3
6a	12	11	10	9	8	7	6	5	4

Early November

	Eastern Hemisphere				Zenith	Western Hemisphere			
	Low			High		High			Low
6p	1	24	23	22	21	20	19	18	17
7p	2	1	24	23	22	21	20	19	18
8p	3	2	1	24	23	22	21	20	19
9p	4	3	2	1	24	23	22	21	20
10p	5	4	3	2	1	24	23	22	21
11p	6	5	4	3	2	1	24	23	22
12a	7	6	5	4	3	2	1	24	23
1a	8	7	6	5	4	3	2	1	24
2a	9	8	7	6	5	4	3	2	1
3a	10	9	8	7	6	5	4	3	2
4a	11	10	9	8	7	6	5	4	3
5a	12	11	10	9	8	7	6	5	4
6a	13	12	11	10	9	8	7	6	5

Late November

	Eastern Hemisphere				Zenith	Western Hemisphere			
	Low			High		High			Low
6p	2	1	24	23	22	21	20	19	18
7p	3	2	1	24	23	22	21	20	19
8p	4	3	2	1	24	23	22	21	20
9p	5	4	3	2	1	24	23	22	21
10p	6	5	4	3	2	1	24	23	22
11p	7	6	5	4	3	2	1	24	23
12a	8	7	6	5	4	3	2	1	24
1a	9	8	7	6	5	4	3	2	1
2a	10	9	8	7	6	5	4	3	2
3a	11	10	9	8	7	6	5	4	3
4a	12	11	10	9	8	7	6	5	4
5a	13	12	11	10	9	8	7	6	5
6a	14	13	12	11	10	9	8	7	6

Early December

	Eastern Hemisphere				Zenith	Western Hemisphere			
	Low			High		High			Low
6p	3	2	1	24	23	22	21	20	19
7p	4	3	2	1	24	23	22	21	20
8p	5	4	3	2	1	24	23	22	21
9p	6	5	4	3	2	1	24	23	22
10p	7	6	5	4	3	2	1	24	23
11p	8	7	6	5	4	3	2	1	24
12a	9	8	7	6	5	4	3	2	1
1a	10	9	8	7	6	5	4	3	2
2a	11	10	9	8	7	6	5	4	3
3a	12	11	10	9	8	7	6	5	4
4a	13	12	11	10	9	8	7	6	5
5a	14	13	12	11	10	9	8	7	6
6a	15	14	13	12	11	10	9	8	7

Late December

	Eastern Hemisphere				Zenith	Western Hemisphere			
	Low			High		High			Low
6p	4	3	2	1	24	23	22	21	20
7p	5	4	3	2	1	24	23	22	21
8p	6	5	4	3	2	1	24	23	22
9p	7	6	5	4	3	2	1	24	23
10p	8	7	6	5	4	3	2	1	24
11p	9	8	7	6	5	4	3	2	1
12a	10	9	8	7	6	5	4	3	2
1a	11	10	9	8	7	6	5	4	3
2a	12	11	10	9	8	7	6	5	4
3a	13	12	11	10	9	8	7	6	5
4a	14	13	12	11	10	9	8	7	6
5a	15	14	13	12	11	10	9	8	7
6a	16	15	14	13	12	11	10	9	8

List 1

Pi And   37 Cet   77 Psc

Multiple Stars

Con.	Object	R.A.	Dec.	Pair	Sep.	Mag.
And	Pi And	00h 37m	+33° 43'	AB	36"	4.3, 7.1
	STF 79	01h 00m	+44° 43'	AB	8"	6.0, 6.8
Cas	Eta Cas	00h 49m	+57° 49'	AB	13"	3.5, 7.4
	Phi Cas	01h 20m	+64° 39'	AC	135"	5.1, 7.0
Cet	37 Cet	01h 14m	-07° 55'	AB	50"	5.2, 7.8
Phe	Zet Phe	01h 08m	-55° 15'	AB	7"	4.0, 8.2
Psc	55 Psc	00h 40m	+21° 26'	AB	7"	5.6, 8.5
	65 Psc	00h 50m	+27° 43'	AB	4"	6.3, 6.3
	Psi¹ Psc	01h 06m	+21° 28'	AB	30"	5.3, 5.5
	77 Psc	01h 06m	+04° 55'	AB	33"	6.4, 7.3
	Zet Psc	01h 14m	+07° 35'	AB	23"	5.2, 6.2
	S 398	01h 28m	+07° 58'	AB	69"	6.3, 8.0
Tuc	Bet Tuc	00h 32m	-62° 57'	AB	27"	4.3, 4.5
	Kap Tuc	01h 16m	-68° 53'	AB	5"	5.0, 7.0
				AC	320"	5.0, 7.9
	HJ 3426	01h 17m	-66° 24'	AB	2"	6.4, 8.3

Open Clusters

Con.	Object	R.A.	Dec.	Diam.	Mag.
Cas	King 14	00h 32m	+63° 10'	7'	8.5
	Ber 3	00h 40m	+61° 58'	4'	8.8
	NGC 225	00h 44m	+61° 47'	12'	7.0
	NGC 436	01h 16m	+58° 49'	5'	8.8
	NGC 457	01h 20m	+58° 17'	13'	6.4
Cep	NGC 188	00h 47m	+85° 14'	13'	8.1

NGC 457   NGC 281

Open Clusters with Nebulosity

Con.	Object	R.A.	Dec.	Diam.	Mag.
Cas	NGC 281	00h 53m	+56° 37'	28'	7.4

Planetary Nebulae

Con.	Object	R.A.	Dec.	Diam.	Mag.
Cet	NGC 246	00h 47m	-11° 52'	240"	10.9

NGC 246   NGC 288

Globular Clusters

Con.	Object	R.A.	Dec.	Diam.	Mag.
Scl	NGC 288	00h 52m	-26° 35'	14'	8.1
Tuc	NGC 104	00h 24m	-72° 05'	50'	4.0

Galaxies

Con.	Object	R.A.	Dec.	Diam.	Mag.	Type
And	M 110	00h 40m	+41° 41'	22'	8.2	Gx E
	M 32	00h 43m	+40° 52'	9'	8.3	Gx E
	M 31	00h 43m	+41° 16'	191'	4.2	Gx S
	NGC 404	01h 09m	+35° 43'	4'	10.2	Gx S
Cas	NGC 147	00h 33m	+48° 31'	13'	9.6	Gx E
	NGC 185	00h 39m	+48° 20'	12'	9.3	Gx E
	NGC 278	00h 52m	+47° 33'	2'	10.9	Gx S
Cet	NGC 157	00h 35m	-08° 24'	4'	10.4	Gx S
	NGC 210	00h 41m	-13° 52'	5'	11.1	Gx S
	NGC 247	00h 48m	-20° 46'	21'	9.2	Gx S
	NGC 337	01h 00m	-07° 35'	3'	11.6	Gx S
	IC 1613	01h 05m	+02° 08'	16'	9.2	Gx ?
	NGC 428	01h 13m	+00° 59'	4'	11.2	Gx S
	NGC 578	01h 30m	-22° 40'	5'	11.1	Gx S
Psc	NGC 488	01h 22m	+05° 15'	5'	10.6	Gx S
	NGC 524	01h 25m	+09° 32'	3'	10.5	Gx S
Scl	NGC 150	00h 34m	-27° 48'	4'	11.4	Gx S
	NGC 253	00h 48m	-25°17'	28'	7.8	Gx S
	NGC 289	00h 53m	-31° 12'	5'	11.0	Gx S
	NGC 300	00h 55m	-37° 41'	22'	8.3	Gx S
Tuc	NGC 292	00h 53m	-72° 48'	316'	2.2	Gx S

M 31

NGC 253

List 2

14 Ari   Iot Cas   Alp Psc

Multiple Stars

Con.	Object	R.A.	Dec.	Pair	Sep.	Mag.
And	56 And	01h 56m	+37° 15'	AB	201"	5.8, 6.1
	Gam And	02h 04m	+42° 20'	AB	10"	2.3, 5.0
	59 And	02h 11m	+39° 02'	AB	17"	6.1, 6.7
Ari	1 Ari	01h 50m	+22° 17'	AB	3"	6.3, 7.2
	Gam Ari	01h 54m	+19° 18'	AB	8"	4.5, 4.6
	Lam Ari	01h 58m	+23° 36'	AB	38"	4.8, 6.7
	14 Ari	02h 09m	+25° 56'	AB	93"	5.0, 8.0
				AC	106"	5.0, 7.5
Cas	Iot Cas	02h 29m	+67° 24'	AB	3"	4.6, 6.9
Cet	66 Cet	02h 13m	-02°24'	AB	17"	5.7, 7.7
Eri	Rho Eri	01h 40m	-56° 12'	AB	11"	5.8, 5.9
Per	STT 25	02h 17m	+57° 03'	AB	103"	6.5, 7.4
Psc	Alp Psc	02h 02m	+02° 46'	AB	2"	4.1, 5.2
Scl	Eps Scl	01h 46m	-25° 03'	AB	5"	5.4, 8.5
Tri	Iot Tri	02h 12m	+30° 18'	AB	4"	5.3, 6.7

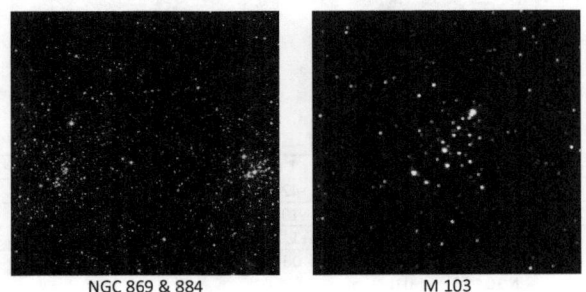

NGC 869 & 884   M 103

Open Clusters

Con.	Object	R.A.	Dec.	Diam.	Mag.
And	NGC 752	01h 58m	+37° 51'	50'	5.7
Cas	M 103	01h 33m	+60° 39'	6'	7.4
	Col 15	01h 36m	+61°17'	5'	8.1
	NGC 637	01h 43m	+64° 02'	4'	8.2
	NGC 654	01h 44m	+61° 53'	5'	6.5
	NGC 659	01h 44m	+60° 40'	5'	7.9
	Col 463	01h 48m	+71° 44'	36'	5.7
	NGC 663	01h 46m	+61° 13'	16'	7.1
	Sto 2	02h 15m	+59° 27'	60'	4.4
Per	NGC 744	01h 58m	+55° 28'	11'	7.9
	NGC 869	02h 19m	+57° 08'	29'	5.3
	NGC 884	02h 23m	+57° 09'	29'	6.1
Tri	Col 21	01h 50m	+27° 04'	9'	8.2

Planetary Nebulae

Con.	Object	R.A.	Dec.	Diam.	Mag.
Per	M 76	01h 42m	+51° 34'	180"	10.1

NGC 663

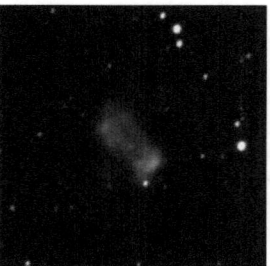

M 76

Galaxies

Con.	Object	R.A.	Dec.	Diam.	Mag.	Type
And	NGC 891	02h 23m	+42° 21'	14'	10.0	Gx S
Ari	NGC 772	01h 59m	+19° 00'	7'	10.4	Gx S
	NGC 821	02h 08m	+11° 00'	3'	11.2	Gx E
Cet	NGC 584	01h 31m	-06° 52'	4'	10.3	Gx E
	NGC 596	01h 33m	-07° 02'	3'	10.8	Gx E
	NGC 720	01h 53m	-13° 44'	5'	10.3	Gx E
	NGC 779	02h 00m	-05° 58'	4'	11.3	Gx S

Galaxies (cont.)

Con.	Object	R.A.	Dec.	Diam.	Mag.	Type
Cet	NGC 864	02h 15m	+06° 00'	5'	11.1	Gx S
	NGC 908	02h 23m	-21° 14'	6'	10.5	Gx S
	NGC 936	02h 28m	-01° 09'	5'	10.0	Gx S
Eri	NGC 685	01h 48m	-52° 46'	4'	11.5	Gx S
Phe	NGC 625	01h 35m	-41° 26'	6'	11.4	Gx S
Psc	M 74	01h 37m	+15° 47'	11'	9.5	Gx S
	NGC 660	01h 43m	+13° 39'	8'	11.4	Gx S
	NGC 676	01h 49m	+05° 54'	4'	11.9	Gx S
Scl	NGC 613	01h 34m	-29° 25'	6'	10.1	Gx S
Tri	M 33	01h 34m	+30° 40'	71'	5.8	Gx S
	NGC 672	01h 48m	+27° 26'	7'	10.8	Gx S
	NGC 925	02h 27m	+33° 35'	11'	9.7	Gx S

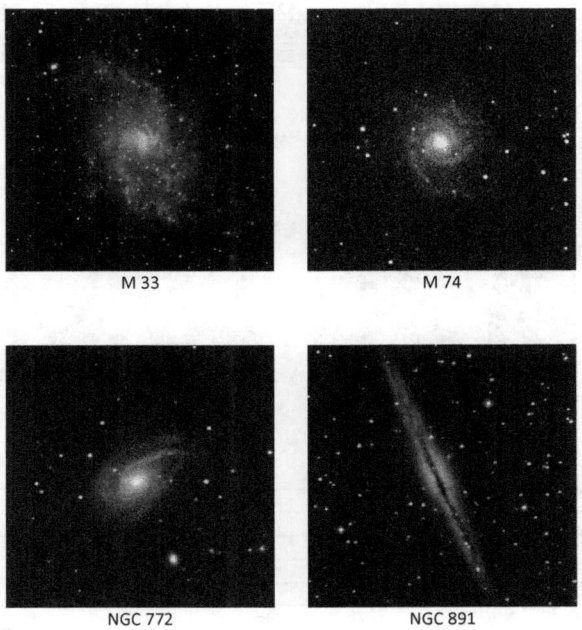

M 33

M 74

NGC 772

NGC 891

List 3

Multiple Stars

Con.	Object	R.A.	Dec.	Pair	Sep.	Mag.
Ari	30 Ari	02h 37m	+24° 39'	AB	39"	6.5, 7.0
	Pi Ari	02h 49m	+17° 28'	AB	3"	5.3, 8.0
Cam	STF 385	03h 29m	+59° 56'	AB	3"	4.2, 7.8
	STF 389	03h 30m	+59° 22'	AB	3"	6.4, 7.9
Cet	Gam Cet	02h 43m	+03° 14'	AB	2"	3.6, 6.2
Eri	The Eri	02h 58m	-40° 18'	AB	8"	3.2, 4.1
For	Ome For	02h 34m	-28° 14'	AB	11"	5.0, 7.7
	Alp For	03h 12m	-28° 59'	AB	5"	4.0, 7.2
Hyi	HJ 3568	03h 08m	-78° 59'	AB	15"	5.7, 7.7
Per	Eta Per	02h 51m	+55° 54'	AB	29"	3.8, 8.5
	STF 331	03h 01m	+52° 21'	AB	12"	5.2, 6.2
Ret	Zeta Ret	03h 18m	-62° 30'	AB	309	5.3, 5.6
Tri	15 Tri	02h 36m	+34°41'	AB	142"	5.4, 6.8
UMi	Alp UMi	02h 32m	+89° 16'	AB	19"	2.1, 9.1

M 34

NGC 1245

Open Clusters

Con.	Object	R.A.	Dec.	Diam.	Mag.
And	NGC 956	02h 33m	+44° 36'	7'	8.9
Cas	Col 33	03h 01m	+60° 27	39'	5.9
	Col 34	03h 01m	+60° 37"	25'	6.8
	Col 36	03h 12m	+63° 11'	23'	7.0
Per	NGC 957	02h 33m	+57° 34'	11'	7.6
	Col 29	02h 37m	+55° 56'	20'	5.9
	M 34	02h 47m	+42° 45'	35'	5.2
	NGC 1245	03h 15m	+47° 14'	10'	8.4

Open Clusters with Nebulae

Con.	Object	R.A.	Dec.	Diam.	Mag.
Cas	IC 1805	02h 33m	+61° 27'	21'	6.5
	NGC 1027	02h 43m	+61° 36'	20'	6.7
	IC 1848	02h 51m	+60° 26'	18'	6.5

IC 1805   NGC 1261

Globular Clusters

Con.	Object	R.A.	Dec.	Diam.	Mag.
Hor	NGC 1261	03h 12m	-55° 13'	7'	8.3

Galaxies

Con.	Object	R.A.	Dec.	Diam.	Mag.	Type
And	IC 239	02h 37m	+38° 58'	5'	11.9	Gx S
Cet	NGC 988	02h 35m	-09° 22'	5'	11.0	Gx S
	NGC 1042	02h 40m	-08° 26'	5'	11.3	Gx S
	NGC 1052	02h 41m	-08° 15'	3'	10.6	Gx E
	NGC 1055	02h 42m	+00° 27'	7'	10.9	Gx S
	M 77	02h 43m	-00° 01'	7'	9.1	Gx S
	NGC 1073	02h 44m	+01° 23'	5'	11.1	Gx S
	NGC 1087	02h 46m	-00° 30'	4'	10.8	Gx S
Eri	NGC 1084	02h 46m	-07° 35'	3'	11.1	Gx S
	NGC 1187	03h 03m	-22° 52'	6'	10.7	Gx S
	NGC 1232	03h 04m	-20° 35'	7'	10.1	Gx S
	NGC 1291	03h 17m	-41° 06'	10'	8.6	Gx S
	NGC 1300	03h 20m	-19° 25'	6'	10.5	Gx S
	NGC 1309	03h 22m	-15° 24'	2'	11.3	Gx S
	NGC 1332	03h 26m	-21° 20'	5'	10.4	Gx S
For	NGC 986	02h 34m	-39° 03'	4'	11.0	Gx S
	NGC 1097	02h 46m	-30° 16'	9'	9.4	Gx S
	NGC 1201	03h 04m	-26° 04'	4'	10.8	Gx S

Galaxies (cont.)

Con.	Object	R.A.	Dec.	Diam.	Mag.	Type
For	NGC 1255	03h 14m	-25° 44'	4'	11.2	Gx S
	NGC 1302	03h 20m	-26° 04'	4'	10.6	Gx S
	NGC 1316	03h 23m	-37° 13'	12'	8.5	Gx S
	NGC 1317	03h 23m	-37° 06'	3'	11.2	Gx S
	NGC 1326	03h 24m	-36° 28'	4'	10.5	Gx S
	NGC 1340	03h 28m	-30° 04'	5'	10.5	Gx E
Per	NGC 1023	02h 40m	+39° 04'	9'	9.5	Gx S
	NGC 1058	02h 43m	+37° 20'	3'	11.4	Gx S
Ret	NGC 1313	03h 18m	-66° 30'	9'	9.2	Gx S

M 77　　　　　　　　　NGC 1232

NGC 1097　　　　　　　NGC 1313

List 4

Multiple Stars

Con.	Object	R.A.	Dec.	Pair	Sep.	Mag.
Cam	STF 396	03h 34m	+58° 46'	AB	20"	6.4, 7.9
Eri	F Eri	03h 49m	-37° 37'	AB	8"	4.7, 5.3
	32 Eri	03h 54m	-02° 57'	AB	7"	4.8, 5.9
Per	Eps Per	03h 58m	+40° 01'	AB	9"	3.0, 7.5
Ret	The Ret	04h 18m	-63° 15'	AB	4"	6.0, 7.7
Tau	STF 7	03h 31m	+27° 44'	AB	43"	7.4, 7.8
	STF 401	03h 31m	+27° 34'	AB	11"	6.6, 6.9
	STF 422	03h 37m	+00° 35'	AB	7"	6.0, 8.9
	19 Tau	03h 45m	+24° 28'	AB	72"	4.3, 8.1
	21 Tau	03h 46m	+24° 32'	AB	168"	5.6, 6.4
	STF 495	04h 08m	+15° 10'	AB	4"	6.1, 8.8
	H 6 98	04h 16m	+06° 11'	AB	64"	6.4, 7.0
	Phi Tau	04h 20m	+27° 21'	AB	49"	5.1, 7.5
	59 Tau	04h 23m	+25°38'	AB	19"	5.4, 8.5
	62 Tau	04h 24m	+24° 18'	AB	29"	6.4, 7.9
	Kap Tau	04h 25m	+22° 18'	AB	340"	4.2, 5.3
	65/67 Tau	04h 25m	+22° 18'	AB	340"	4.2, 5.3
	The Tau	04h 29m	+15°52'	AB	337"	3.4, 3.9
	STF 548	04h 29m	+30°22'	AB	15"	6.4, 8.0

Open Clusters

Con.	Object	R.A.	Dec.	Diam.	Mag.
Cam	Tom 5	03h 48m	+59° 04'	17'	8.4
	NGC 1502	04h 08m	+62° 20'	7'	6.9
Per	NGC 1342	03h 32m	+37° 22'	14'	6.7
	NGC 1444	03h 49m	+52° 40'	4'	6.6
	NGC 1513	04h 10m	+49° 31'	9'	8.4
	NGC 1528	04h 16m	+51° 13'	23'	6.4
	NGC 1545	04h 21m	+50° 15'	18'	6.2
Tau	Mel 25	04h 27m	+16° 00'	330'	0.5

M 45   NGC 1528

Open Clusters with Nebulae

Con.	Object	R.A.	Dec.	Diam.	Mag.
Per	IC 348	03h 45m	+32° 17′	7′	7.3
Tau	M 45	03h 47m	+24° 07′	110′	1.6

Nebulae

Con.	Object	R.A.	Dec.	Diam.	Mag.
Per	NGC 1499	04h 03m	+36° 22′	160′	5.0

Planetary Nebulae

Con.	Object	R.A.	Dec.	Diam.	Mag.
Eri	NGC 1535	04h 14m	-12° 44′	45″	9.6
For	NGC 1360	03h 33m	-25° 52′	9″	9.4
Tau	NGC 1514	04h 09m	+30° 47′	2′	10.9

NGC 1365   NGC 1566

Galaxies

Con.	Object	R.A.	Dec.	Diam.	Mag.	Type
Cam	IC 342	03h 47m	+68° 06'	18'	9.6	Gx S
	IC 365	04h 09m	+69° 51'	4'	11.5	Gx S
Dor	NGC 1515	04h 04m	-54° 06'	5'	11.4	Gx S
	NGC 1533	04h 10m	-56° 07'	3'	10.9	Gx S
	NGC 1549	04h 16m	-55° 36'	5'	9.7	Gx E
	NGC 1553	04h 16m	-55° 47'	5'	9.4	Gx S
	NGC 1566	04h 20m	-54° 56'	7'	9.8	Gx S
	NGC 1596	04h 28m	-55° 02'	4'	11.2	Gx S
Eri	NGC 1395	03h 38m	-23° 02'	6'	10.0	Gx E
	NGC 1404	03h 39m	-35° 36'	3'	10.2	Gx E
	NGC 1400	03h 40m	-18° 41'	2'	11.3	Gx E
	NGC 1407	03h 40m	-18°35'	5'	10.1	Gx E
	NGC 1421	03h 42m	-13° 29'	4'	11.6	Gx S
	NGC 1532	04h 12m	-32° 53'	11'	10.6	Gx S
	NGC 1537	04h 14m	-31° 38'	4'	10.6	Gx E
For	NGC 1350	03h 31m	-33° 38'	5'	10.7	Gx S
	NGC 1365	03h 34m	-36° 08'	11'	9.6	Gx S
	NGC 1367	03h 35m	-24° 56'	9'	10.8	Gx S
	NGC 1374	03h 35m	-35° 14'	3'	11.2	Gx E
	NGC 1379	03h 36m	-35° 26'	2'	11.2	Gx E
	NCG 1380	03h 36m	-34° 59'	5'	10.2	Gx S
	NGC 1387	03h 37m	-35° 30'	3'	10.9	Gx S
	NGC 1385	03h 37m	-24° 30'	3'	11.1	Gx S
	NGC 1399	03h 38m	-35° 27'	7'	9.4	Gx E
	NGC 1398	03h 39m	-26° 20'	7'	9.8	Gx S
	NGC 1425	03h 42m	-29° 54'	6'	11.0	Gx S
	NGC 1427	03h 42m	-35° 24'	4'	11.1	Gx S
Hor	NGC 1433	03h 42m	-47° 13'	7'	10.1	Gx S
	NGC 1448	03h 45m	-44° 39'	8'	11.0	Gx S
	NGC 1493	03h 57m	-46° 13'	4'	11.4	Gx S
	NGC 1512	04h 04m	-43° 21'	9'	10.6	Gx S
	NGC 1527	04h 08m	-47° 54'	4'	11.0	Gx S
Hyi	NGC 1511	04h 00m	-67° 38'	5'	11.4	Gx I
Ret	NGC 1543	04h 13m	-57° 44'	5'	10.3	Gx S
	NGC 1559	04h 18m	-62° 47'	4'	10.7	Gx S
	NGC 1574	04h 22m	-56° 58'	3'	10.7	Gx S

List 5

Multiple Stars

Con.	Object	R.A.	Dec.	Pair	Sep.	Mag.
Aur	Ome Aur	04h 59m	+37° 53'	AB	5"	5.1, 8.1
	14 Aur	05h 15m	+32° 41'	AC	14"	5.0, 7.3
Cam	1 Cam	04h 32m	+53° 55'	AB	11"	5.8, 6.8
	Bet Cam	05h 03m	+60° 27'	AB	83"	4.1, 7.4
	11 Cam	05h 06m	+58° 58'	AB	179"	5.2, 6.2
Eri	STF 570	04h 35m	-09° 44'	AB	13"	6.7, 7.6
	55 Eri	04h 44m	-08° 48'	AB	9"	6.7, 6.8
Lep	Kap Lep	05h 13m	-12° 56'	AB	2"	4.5, 6.8
	S 476	05h 19m	-18° 31'	AB	40"	6.3, 6.5
	HJ 3752	05h 22m	-24° 46'	AB	3"	5.4, 6.6
	HJ 3759	05h 26m	-19° 42'	AB	27"	5.9, 7.3
	Bet Lep	05h 28m	-20° 46'	AB	2"	3.0, 7.5
Ori	SHJ 49	04h 59m	+14° 33'	AB	40"	6.1, 7.4
	STF 630	05h 02m	+01° 37'	AC	14"	6.2, 7.7
	Bet Ori	05h 15m	-08° 12'	AB	9"	0.3, 6.8
	23 Ori	05h 23m	+03° 33'	AB	32"	5.0, 6.8
Per	57 Per	04h 33m	+43° 04'	AB	121"	6.1, 6.8
Pic	Iot Pic	04h 51m	-53° 28'	AB	13"	5.6, 6.2
	The Pic	05h 25m	-52° 19'	AC	38"	6.2, 6.7
	DUN 21	05h 30m	-47° 05'	AD	198"	5.5, 6.7
Tau	88 Tau	04h 36m	+10° 10'	AB	69"	4.3, 7.8
	Rho Tau	04h 39m	+15° 55'	AB	431"	4.7, 5.1
	Tau Tau	04h 42m	+22° 57'	AB	63"	4.2, 7.0
	118 Tau	05h 29m	+25° 09'	AB	5"	5.8, 6.7

Theta Pic   M 38

Open Clusters

Con.	Object	R.A.	Dec.	Diam.	Mag.
Aur	NGC 1664	04h 51m	+43°40'	18'	7.6
	NGC 1778	05h 08m	+37° 01'	6'	7.7
	NGC 1857	05h 20m	+39° 21'	5'	7.0
	Col 62	05h 22m	+41° 00'	28'	4.2
	M 38	05h 29m	+35° 51'	21'	6.4
Cam	Col 464	05h 22m	+73° 17'	120'	4.2
Ori	NGC 1662	04h 48m	+10° 56'	20'	6.4
	Col 65	05h 26m	+16° 06'	220'	3.0
Per	NGC 1582	04h 32m	+43° 51'	37'	7.0
Tau	NGC 1647	04h 46m	+19° 07'	45'	6.4
	NGC 1746	05h 04m	+23° 46'	42'	6.1
	NGC 1807	05h 11m	+16° 32'	17'	7.0
	NGC 1817	05h 12m	+16° 41'	16'	7.7

Open Clusters with Nebulae

Con.	Object	R.A.	Dec.	Diam.	Mag.
Aur	NGC 1893	05h 22m	+33° 25'	11'	7.5

IC 405   M 79

Nebulae

Con.	Object	R.A.	Dec.	Diam.	Mag.
Aur	IC 405	05h 16m	+34° 16'	50'	10.0
Dor	NGC 1966	05h 27m	-68° 49'	13'	8.5

Planetary Nebulae

Con.	Object	R.A.	Dec.	Diam.	Mag.
Lep	IC 418	05h 27m	-12° 42'	13"	9.3

Globular Clusters

Con.	Object	R.A.	Dec.	Diam.	Mag.
Col	NGC 1851	05h 14m	-40° 03'	11'	7.1
Lep	M 79	05h 24m	-24° 31'	9'	7.7

NGC 1851

NGC 1808

Galaxies

Con.	Object	R.A.	Dec.	Diam.	Mag.	Type
Col	NGC 1792	05h 05m	-37° 59'	5'	10.2	Gx S
	NGC 1808	05h 08m	-37° 31'	7'	10.0	Gx S
Dor	NGC 1617	04h 32m	-54° 36'	4'	10.7	Gx S
	NGC 1672	04h 46m	-59° 15'	7'	10.2	Gx S
	NGC 1703	04h 54m	-53° 22'	2'	11.9	Gx S
	LMC	05h 24m	-69° 45'	645'	0.9	Gx S
	NGC 1947	05h 27m	-63° 46'	3'	10.9	Gx S
Eri	NGC 1600	04h 32m	-05° 05'	3'	11.5	Gx E
	NGC 1637	04h 41m	-02° 51'	4'	11.1	Gx S
	NGC 1700	04h 57m	-04° 52'	3'	11.1	Gx E
Lep	NGC 1744	05h 00m	-26° 01'	8'	11.5	Gx S
	NGC 1832	05h 12m	-15° 41'	3'	11.8	Gx S

List 6

Multiple Stars

Con.	Object	R.A.	Dec.	Pair	Sep.	Mag.
Aur	STT 63	05h 31m	+39° 50'	AB	76"	6.5, 7.7
	STF 764	05h 39m	+29° 29'	AB	26"	6.4, 7.1
	The Aur	06h 00m	+37° 13'	AB	4"	2.7, 7.2
	41 Aur	06h 12m	+48° 43'	AB	8"	6.2, 6.9
CMa	Zet CMa	06h 20m	-30° 04'	AB	176"	3.0, 7.7
Col	HJ 3857	06h 24m	-36° 42'	AB	64"	5.7, 6.9
	HJ 3858	06h 26m	-35° 04'	AB	132"	6.4, 7.6
Gem	18 Gem	06h 28m	+20° 13'	AB	112"	4.1, 8.0
Lep	Gam Lep	05h 45m	-22° 27'	AB	97"	3.6, 6.3
Lyn	5 Lyn	06h 27m	+58° 25'	AC	96"	5.4, 7.9
Mon	Eps Mon	06h 24m	+04° 36'	AB	12"	4.4, 6.6
	Bet Mon	06h 29m	-07° 02'	AB	7"	4.6, 5.0
				BC	3"	5.0, 5.3
Ori	Del Ori	05h 32m	-00° 18'	AB	53"	2.4, 6.8
	Lam Ori	05h 35m	+09° 56'	AB	4"	3.5, 5.5
	The¹ Ori	05h 35m	-05° 25'	AB	135"	5.0, 5.1
	The² Ori	05h 35m	-05° 25'	AB	52"	5.0, 6.2
	Iot Ori	05h 35m	-05° 55'	AB	11"	2.9, 7.0
	STF 747	05h 35m	-06° 00'	AB	36"	4.7, 5.5
	Rho Ori	05h 39m	-02° 36'	AD	13"	3.8, 6.6
				AE	42"	3.8, 6.3
	Zet Ori	05h 41m	01° 57'	AB	2"	1.9, 4.0
	STF 855	06h 09m	+02° 30'	AB	29"	5.7, 6.7
				AC	119"	5.7, 9.7
Pic	DUN 27	06h 16m	-59° 13'	AB	34"	6.5, 7.6
Pup	HJ 3834	06h 05m	-45° 05'	AC	196"	6.0, 6.4
	HR 2384	06h 30m	-50° 14'	AC	12"	6.0, 8.0
Tau	STF 730	05h 32m	+17°03'	AB	10"	6.1, 6.4

HJ 3857

DUN 27

M 36   M 37

Open Clusters

Con.	Object	R.A.	Dec.	Diam	Mag.
Aur	M 36	05h 36m	+34° 08'	12'	6.0
	M 37	05h 52m	+32° 33'	23'	5.6
CMa	NGC 2204	06h 16m	-18° 40'	12'	8.6
Gem	NGC 2129	06h 01m	+23° 19'	6'	6.7
	M 35	06h 09m	+24° 21'	28'	5.1
Mon	NGC 2215	06h 21m	-07° 17'	11'	8.4
	Col 91	06h 22m	+02° 19'	13'	6.4
	Col 92	06h 23m	+05° 07'	11'	8.6
	NGC 2232	06h 28m	-04° 51'	29'	4.2
	Col 96	06h 30m	+02° 52'	7'	7.3
Ori	NGC 2169	06h 08m	+13° 58'	6'	5.9
	NGC 2186	06h 12m	+05° 28'	4'	8.7
	NGC 2194	06h 14m	+12° 48'	10'	8.5

M 35   NGC 2169

Open Clusters with Nebulae

Con.	Object	R.A.	Dec.	Diam.	Mag.
Gem	Col 89	06h 18m	+23° 38'	35'	5.7
Ori	Col 69	05h 35m	+09° 53'	64'	2.8
	NGC 1981	05h 35m	-04° 26'	25'	4.2
	NGC 1977	05h 35m	-04° 51'	40'	?
	NGC 1980	05h 35m	-05° 55'	13'	2.5
	Col 70	05h 36m	-01° 00'	150'	0.4
	NGC 2175	06h 09m	+20° 29'	40'	6.8

NGC 1977   NGC 2070

Nebulae

Con.	Object	R.A.	Dec.	Diam	Mag.
Dor	NGC 2070	05h 39m	-69° 06'	40'	5.0
	NGC 2074	05h 39m	-69° 30'	16'	8.5
Ori	NGC 1975	05h 35m	-04° 41'	10'	7.0
	M 43	05h 35m	-05° 16'	20'	9.0
	M 42	05h 35m	-05° 23'	90'	4.0
	M 78	05h 47m	+00° 05'	8'	8.0
	NGC 2071	05h 47m	+00° 18'	7'	8.0

Planetary Nebulae

Con.	Object	R.A.	Dec.	Diam	Mag.
Aur	IC 2149	05h 56m	+46° 06'	13"	10.7
CMa	IC 2165	06h 22m	-12° 59'	8"	10.6

M 42 & M 43

M 78

Supernova Remnants

Con.	Object	R.A.	Dec.	Diam	Mag.
Tau	M 1	05h 35m	+22° 01′	5′	8.4

M 1

NGC 2146

Galaxies

Con.	Object	R.A.	Dec.	Diam	Mag.	Type
Cam	NGC 1961	05h 42m	+69° 23′	5′	11.3	Gx S
	NGC 2146	06h 19m	+78° 21′	6′	10.3	Gx S
CMa	IC 2163	06h 17m	-21° 23′	3′	11.8	Gx S
	NGC 2217	06h 22m	-27° 14′	5′	10.4	Gx S
Col	NGC 2090	05h 47m	-34° 15′	5′	11.3	Gx S
Lep	NGC 1964	05h 33m	-21° 57′	6′	10.9	Gx S
	NGC 2139	06h 01m	-23° 40′	3′	11.7	Gx S
	NGC 2196	06h 12m	-21° 48′	3′	11.5	Gx S

List 7

Multiple Stars

Con.	Object	R.A.	Dec.	Pair	Sep.	Mag.
CMa	Nu¹ CMa	06h 36m	-18° 40'	AB	18"	5.8, 7.7
	Mu CMa	06h 56m	-14° 03'	AB	3"	5.3, 7.1
	Eps CMa	06h 59m	-28° 58'	AB	7"	1.5, 7.5
	Del CMa	07h 08m	-26° 24'	AB	266"	1.8, 7.7
	145 CMa	07h 17m	-23° 19'	AB	26"	5.0, 5.8
	BSO 2	07h 17m	-30° 54'	AB	38"	6.3, 7.8
	Eta CMa	07h 24m	-29° 18'	AB	179"	2.5, 6.8
	DUN 47	07h 25m	-31° 49'	AC	99"	5.4, 7.6
Car	HR 2814	07h 20m	-52° 19'	AB	9"	6.0, 6.5
Gem	20 Gem	06h 32m	+17° 47'	AB	20"	6.3, 6.9
	38 Gem	06h 55m	+13° 11'	AB	7"	4.7, 7.7
Lyn	12 Lyn	06h 46m	+59° 27'	AC	9"	5.4, 7.0
	STF 958	06h 48m	+55° 42'	AB	5"	6.3, 6.3
	19 Lyn	07h 23m	+55° 17'	AB	15"	5.8, 6.7
				AD	215"	5.8, 7.6
Mon	15 Mon	06h 41m	+09° 54'	AB	3"	4.6, 7.8
	STT 82	07h 04m	+01° 29'	AB	90"	6.5, 7.6
Pup	DUN 31	06h 39m	-48° 13'	AB	13"	5.1, 7.4
	DUN 38	07h 04m	-46° 37'	AB	21"	5.6, 6.7
	HJ 3928	07h 06m	-34° 47'	AB	3"	6.5, 7.8
	NW Pup	07h 18m	-36° 44'	AB	240"	4.7, 5.1
	DUN 49	07h 29m	-31° 51'	AB	9"	6.3, 7.0
	STF 1104	07h 29m	-15° 00'	AB	2"	6.4, 7.6
Vol	Gam Vol	07h 09m	-70° 30'	AB	14"	3.9, 5.4

M 41

NGC 2362

Open Clusters

Con.	Object	R.A.	Dec.	Diam	Mag.
Aur	NGC 2281	06h 48m	+41° 05'	14'	5.4
CMa	M 41	06h 46m	-20° 45'	38'	4.5
	Col 121	06h 54m	-24° 25'	50'	2.6
	NGC 2345	07h 08m	-13° 12'	12'	7.7
	NGC 2354	07h 15m	-25° 42'	20'	6.5
	Col 132	07h 14m	-31° 10'	95'	3.6
	NGC 2360	07h 18m	-15° 38'	12'	7.2
	NGC 2362	07h 19m	-24° 57'	8'	3.8
	NGC 2367	07h 20m	-21° 53'	4'	7.9
	NGC 2374	07h 24m	-13° 16'	19'	8.0
	Col 140	07h 24m	-32° 12'	42'	3.5
	NGC 2383	07h 25m	-20° 57'	5'	8.4
	NGC 2384	07h 25m	-21° 01'	3'	7.4
	Col 145	07h 26m	-24° 13'	6'	10
Gem	NGC 2331	07h 07m	+27° 16'	18'	8.5
	NGC 2395	07h 27m	+13° 36'	12'	8.0
Mon	Col 96	06h 30m	+02° 52'	7'	7.3
	Col 97	06h 31m	+05° 55'	21'	5.4
	NGC 2250	06h 34m	-05° 00'	7'	8.9
	NGC 2251	06h 35m	+08°22'	10'	7.3
	Col 106	06h 37m	+05° 57'	45'	4.6
	Col 107	06h 38m	+04° 44'	35'	5.1
	Col 111	06h 38m	+06° 54'	3'	7.0
	NGC 2286	06h 48m	-03° 09'	14'	7.5
	NGC 2301	06h 52m	+00° 28'	12'	6.0
	NGC 2302	06h 52m	-07° 05'	3'	8.9
	M 50	07h 03m	-08° 23'	16'	5.9
	NGC 2324	07h 04m	+01° 03'	7'	8.4
	NGC 2353	07h 15m	-10° 16'	20'	7.1
Pup	Col 135	07h 17m	-36° 50'	50'	2.1
	NGC 2396	07h 28m	-11° 43'	10'	7.4

Open Clusters with Nebulae

Con.	Object	R.A.	Dec.	Diam	Mag.
Mon	NGC 2244	06h 32m	+04° 57'	24'	4.8
	NGC 2252	06h 35m	+05° 22'	20'	7.7
	NGC 2264	06h 41m	+09° 54'	20'	4.1
	NGC 2335	07h 07m	-10° 02'	12'	7.2
	NGC 2343	07h 08m	-10° 37'	6'	6.7
Pup	Col 146	07h 27m	-23° 57'	5'	7.9

Nebulae

Con.	Object	R.A.	Dec.	Diam	Mag.
Mon	NGC 2238	06h 31m	+05° 01'	80'	6.0
	NGC 2237	06h 31m	+05° 03'	90'	5.5

Planetary Nebulae

Con.	Object	R.A.	Dec.	Diam	Mag.
Gem	NGC 2392	07h 29m	+20° 55'	45"	9.1

M 50   NGC 2392

Globular Clusters

Con.	Object	R.A.	Dec.	Diam	Mag.
Pup	NGC 2298	06h 49m	-36° 00'	7'	9.3

Galaxies

Con.	Object	R.A.	Dec.	Diam	Mag.	Type
Cam	NGC 2336	07h 27m	+80° 11'	7'	10.7	Gx S
	NGC 2366	07h 29m	+69° 11'	8'	11.0	Gx S
CMa	NGC 2280	06h 45m	-27° 38'	6'	11.0	Gx S
	NGC 2292	06h 48m	-26° 45'	4'	10.9	GX S
Cep	NGC 2276	07h 27m	+85° 45'	3'	11.8	Gx S
Pup	NGC 2310	06h 54m	-40° 52'	4'	11.5	Gx S

List 8

Multiple Stars

Con.	Object	R.A.	Dec.	Pair	Sep.	Mag.
Cam	SHJ 86	08h 03m	+63° 05'	AB	51"	6.2, 7.5
Cnc	Zet Cnc	08h 12m	+17° 39'	AC	6"	5.1, 5.9
	Phi² Cnc	08h 27m	+26° 56'	AB	5"	6.2, 6.2
Car	DUN 60	08h 01m	-54° 31'	AB	40"	6.1, 7.9
	RMK 8	08h 15m	-62° 55'	AB	4"	5.3, 7.6
Gem	Alp Gem	07h 35m	+31° 53'	AB	4"	1.9, 3.0
Mon	STF 1183	08h 07m	-09° 15'	AB	31"	6.2, 7.8
Pup	H N 19	07h 34m	-23° 28'	AB	10"	5.8, 5.9
	Kap Pup	07h 39m	-26° 48'	AB	10"	4.4, 4.6
	2 Pup	07h 46m	-14° 41'	AB	17"	6.0, 6.7
	DUN 59	07h 59m	-49° 59'	AB	17"	6.2, 6.2
	DUN 67	08h 14m	-36° 19'	AB	67"	5.0, 6.0
	S 568	08h 25m	-24° 03'	AB	42"	5.5, 8.4
	HJ 4093	08h 26m	-39° 04'	AB	8"	6.5, 7.1
Vel	Gam Vel	08h 10m	-47° 20'	AB	41"	1.8, 4.1
				AC	3"	1.8, 7.3
				AD	94"	1.8, 9.4
	HJ 4069	08h 14m	-45° 50'	AC	32"	5.8, 8.7
	A Vel	08h 29m	-47° 56'	AB	3"	5.5, 7.2
				AC	19"	5.5, 9.2
	DUN 70	08h 30m	-44° 43'	AB	4"	5.2, 7.0
Vol	Eps Vol	08h 08m	-68° 37'	AB	6"	4.4, 7.3
	Kap Vol	08h 20m	-71° 31'	AB	65"	5.3, 5.6
				BC	38"	5.6, 7.7

M 46   M 47

Open Clusters

Con.	Object	R.A.	Dec.	Diam	Mag.
Car	NGC 2516	07h 58m	-60° 45'	29'	3.8
Gem	NGC 2420	07h 38m	+21° 34'	10'	8.3
Hya	M 48	08h 14m	-05° 45'	54'	5.8
Pup	NGC 2414	07h 33m	-15° 27'	4'	7.9
	NGC 2421	07h 36m	-20° 37'	10'	8.3
	M 47	07h 37m	-14° 29'	29'	4.4
	NGC 2423	07h 37m	-13° 52'	19'	6.7
	Mel 71	07h 38m	-12° 04'	9'	7.1
	NGC 2439	07h 40m	-31° 42'	10'	6.9
	M 46	07h 42m	-14° 49'	27'	6.1
	M 93	07h 44m	-23° 51'	22'	6.2
	NGC 2451	07h 45m	-37° 58'	45'	2.8
	NGC 2453	07h 48m	-27° 12'	5'	8.3
	NGC 2477	07h 52m	-38° 32'	27'	5.8
	NGC 2482	07h 55m	-24° 15'	12'	7.3
	NGC 2483	07h 56m	-27° 53'	10'	7.6
	Col 168	07h 56m	-25° 53'	5'	8.7
	NGC 2489	07h 56m	-30° 04'	8'	7.9
	Col 173	08h 04m	-46° 00'	370'	0.6
	NGC 2527	08h 05m	-28° 09'	22'	6.5
	NGC 2533	08h 07m	-29° 53'	4'	7.6
	NGC 2539	08h 11m	-12° 49'	21'	6.5
	NGC 2546	08h 12m	-37° 36'	40'	6.3
	NGC 2567	08h 18m	-30° 39'	10'	7.4
	NGC 2571	08h 19m	-29° 45'	13'	7.0
	NGC 2579	08h 21m	-36° 13'	10'	7.5
	Col 185	08h 23m	-36° 20'	9'	7.8

Open Clusters with Nebulae

Con.	Object	R.A.	Dec.	Diam	Mag.
Pup	NGC 2467	07h 52m	-26° 26'	15'	7.1
Vel	NGC 2547	08h 10m	-49° 14'	20'	4.7

Globular Clusters

Con.	Object	R.A.	Dec.	Diam	Mag.
Lyn	NGC 2419	07h 38m	+38° 53'	4'	10.3

M 93

NGC 2403

Galaxies

Con.	Object	R.A.	Dec.	Diam	Mag.	Type
Cam	NGC 2403	07h 37m	+65° 36'	18'	8.6	Gx S
Pup	NGC 2559	08h 17m	-27° 27'	4'	11.1	Gx S
	NGC 2566	08h 19m	-25° 30'	3'	11.2	Gx S
Vol	NGC 2442	07h 36m	-69° 33'	6'	10.5	Gx S

List 9

Multiple Stars

Con.	Object	R.A.	Dec.	Pair	Sep.	Mag.
Cnc	STF 1245	08h 36m	+06° 37'	AB	10"	6.0, 7.3
	39/40 Cnc	08h 40m	+19° 33'	AB	150"	6.5, 6.6
	Iot Cnc	08h 47m	+28° 46'	AB	31"	4.1, 6.0
Car	RMK 9	08h 45m	-58° 43'	AB	4"	6.9, 6.9
	DUN 74	08h 57m	-59° 14'	AB	40"	4.9, 6.6
Hya	Eps Hya	08h 47m	+06° 25'	AB	3"	3.5, 6.7
	S 585	08h 55m	-18° 14'	AB	64"	5.9, 7.2
	27 Hya	09h 21m	-09° 33'	AB	229"	4.9, 7.0
	Tau^1 Hya	09h 29m	-02° 46'	AB	66"	4.9, 7.3
Lyn	38 Lyn	09h 19m	+36° 48'	AB	3"	3.9, 6.6
UMa	Sig^2 UMa	09h 10m	+67° 08'	AB	4"	4.9, 7.9
	41 UMa	09h 29m	+45° 36'	AB	72"	5.5, 7.8
Vel	HJ 4107	08h 31m	-39° 04'	AB	4"	6.5, 8.2
				AC	30"	6.5, 9.1
	HJ 4126	08h 40m	-53° 03'	AB	17"	5.2, 8.7
	BSO 18	08h 42m	-53° 07'	AB	76"	4.8, 5.5
				BD	60"	5.5, 9.9
	H Vel	08h 56m	-52° 43'	AB	3"	4.7, 7.7
	HJ 4188	09h 13m	-43° 37'	AB	3"	6.0, 6.8

Open Clusters

Con.	Object	R.A.	Dec.	Diam	Mag.
Cnc	M 44	08h 40m	+19° 40'	95'	3.1
	M 67	08h 51m	+11° 49'	29'	6.9
Pyx	NGC 2627	08h 37m	-29° 57'	11'	8.4
Vel	NGC 2645	08h 39m	-46° 14'	3'	7.0
	IC 2391	08h 40m	-53° 04'	50'	2.6
	IC 2395	08h 40m	-48° 07'	7'	4.6
	NGC 2659	08h 43m	-45° 00'	3'	8.6
	NGC 2670	08h 45m	-48° 48'	9'	7.8
	NGC 2669	08h 46m	-52° 57'	12'	6.1
	Col 203	08h 48m	-42° 31'	29'	4.6
	IC 2488	09h 28m	-56° 59'	14'	7.4

 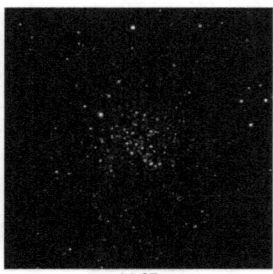

M 44   M 67

Open Clusters with Nebulae

Con.	Object	R.A.	Dec.	Diam	Mag.
Vel	Col 197	08h 44m	-41° 22'	17'	6.7

Planetary Nebulae

Con.	Object	R.A.	Dec.	Diam	Mag.
Car	NGC 2867	09h 21m	-58° 19'	12"	9.7
Pyx	He 2-12	08h 41m	-32° 23'	19'	11.0

Globular Clusters

Con.	Object	R.A.	Dec.	Diam	Mag.
Car	NGC 2808	09h 12m	-64° 52'	14'	6.2

Galaxies

Con.	Object	R.A.	Dec.	Diam	Mag.	Type
Cam	NGC 2715	09h 08m	+78° 05'	5'	11.3	Gx S
Cnc	NGC 2775	09h 10m	+07° 02'	4'	10.4	Gx S
Car	NGC 2822	09h 14m	-69° 39'	3'	10.8	Gx S
Hya	NGC 2784	09h 12m	-24° 10'	6'	10.3	Gx S
	NGC 2835	09h 18m	-22° 21'	7'	10.6	Gx S
LMi	NGC 2859	09h 24m	+34° 31'	4'	10.6	Gx S
Lyn	NGC 2683	08h 53m	+33° 25'	9'	9.6	Gx S
Pyx	NGC 2613	08h 33m	-22° 58'	7'	10.6	Gx S
	NGC 2663	08h 45m	-33° 48'	4'	11.0	Gx E
UMa	NGC 2681	08h 54m	+51° 19'	4'	10.1	Gx S
	NGC 2768	09h 12m	+60° 02'	8'	10.1	Gx E
	NGC 2805	09h 20m	+64° 06'	6'	11.0	Gx S
	NGC 2841	09h 22m	+50° 59'	8'	9.5	Gx S

List 10

Multiple Stars

Con.	Object	R.A.	Dec.	Pair	Sep.	Mag.
Ant	Zeta[1] Ant	09h 31m	-31° 53'	AB	8"	6.2, 6.8
Car	Ups Car	09h 47m	-65° 04'	AB	5"	3.0, 6.0
	HJ 4306	10h 19m	-64° 41'	AB	2"	6.5, 6.3
Hya	SHJ 110	10h 04m	-18° 06'	AB	21"	6.2, 7.0
Leo	Zet Leo	10h 17m	+23° 25'	AB	326"	3.4, 5.9
	Gam Leo	10h 20m	+19° 50'	AB	5"	2.4, 3.6
Sex	HJ 2530	10h 24m	+02° 22'	AB	202"	6.2, 7.1
UMa	STF 1415	10h 18m	+71° 04'	AB	17"	6.7, 7.3
Vel	HJ 4420	09h 34m	-49° 00'	AB	2"	5.5, 6.2
	DUN 81	09h 54m	-45° 17'	AB	5"	5.8, 8.2
	J Vel	10h 21m	-56° 03'	AB	7"	4.5, 7.2
				AC	36"	4.5, 9.2

Open Clusters

Con.	Object	R.A.	Dec.	Diam	Mag.
Car	NGC 3114	10h 03m	-60° 07'	35'	4.2
Vel	NGC 2925	09h 33m	-53° 24'	12'	8.3
	NGC 3033	09h 49m	-56° 25'	5'	8.8
	NGC 3228	10h 21m	-51° 44'	5'	6.0

Open Clusters with Nebulae

Con.	Object	R.A.	Dec.	Diam	Mag.
Car	IC 2581	10h 27m	-57° 38'	7'	4.3

NGC 3242

NGC 3132

Planetary Nebulae

Con.	Object	R.A.	Dec.	Diam	Mag.
Hya	NGC 3242	10h 25m	-18° 39'	41"	7.7
Vel	NGC 3132	10h 07m	-40° 26'	84"	9.2

Globular Clusters

Con.	Object	R.A.	Dec.	Diam	Mag.
Vel	NGC 3201	10h 18m	-46° 25'	18'	6.9

M 81   M 82

Galaxies

Con.	Object	R.A.	Dec.	Diam	Mag.	Type
Ant	NGC 2997	09h 46m	-31° 11'	9'	9.7	Gx S
	NGC 3100	10h 01m	-31° 40'	3'	11.3	Gx S
	NGC 3223	10h 22m	-34° 16'	4'	11.2	Gx S
Car	NGC 3059	09h 50m	-73° 55'	4'	11.1	Gx S
	NGC 3136	10h 06m	-67° 23'	3'	11.0	Gx E
Dra	NGC 3147	10h 17m	+73° 24'	4'	10.8	Gx S
Hya	NGC 2986	09h 44m	-21° 17'	3'	10.9	Gx E
	NGC 3109	10h 03m	-26° 10'	19'	9.9	Gx S
Leo	Leo I	10h 08m	+12° 18'	10'	11.1	Gx E
	NGC 3190	10h 18m	+21° 50'	4'	11.1	Gx S
	NGC 3193	10h 18m	+21° 54'	3'	11.2	Gx E
	NGC 3227	10h 24m	+19° 52'	5'	10.9	Gx S
	NGC 3239	10h 25m	+17° 10'	5'	11.9	Gx I
LMi	NGC 3245	10h 27m	+28° 30'	3'	10.8	Gx S
Sex	NGC 2974	09h 43m	-03° 42'	4'	11.0	Gx E
	NGC 3115	10h 05m	-07° 43'	7'	8.6	Gx E
	NGC 3166	10h 14m	+03° 26'	5'	10.4	Gx S
	NGC 3169	10h 14m	+03° 28'	4'	10.7	Gx S
UMa	NGC 2950	09h 43m	+58° 51'	3'	11.0	Gx S

Galaxies (cont.)

Con.	Object	R.A.	Dec.	Diam	Mag.	Type
UMa	NGC 2976	09h 47m	+67° 55′	6′	10.6	GX S
	NGC 2985	09h 50m	+72° 17′	5′	10.5	Gx S
	M 81	09h 56m	+69° 04′	27′	7.3	Gx S
	M 82	09h 56m	+69° 41′	11′	8.9	Gx I
	NGC 3079	10h 02m	+55° 41′	8′	10.8	Gx S
	NGC 3065	10h 02m	+72° 10′	2′	11.9	Gx S
	NGC 3077	10h 03m	+68° 44′	5′	10.5	Gx I
	NGC 3184	10h 18m	+41° 25′	7′	9.9	Gx S
	NGC 3198	10h 20m	+45° 33′	9′	10.2	Gx S
	IC 2574	10h 28m	+68° 25′	13′	10.4	Gx S

List 11

Multiple Stars

Con.	Object	R.A.	Dec.	Pair	Sep.	Mag.
Car	DUN 94	10h 39m	-59° 11'	AB	15"	4.9, 7.5
	Gliese 152	10h 39m	-58° 49'	AB	26"	6.2, 8.0
	DUN 99	10h 44m	-70° 52'	AB	63"	6.3, 6.5
	u Car	10h 54m	-58° 51'	AB	159"	3.9, 6.5
				AC	56"	3.9, 7.8
Cen	BSO 6	11h 29m	-42° 40'	AB	13"	5.1, 7.4
Crt	Gam Crt	11h 25m	-17° 41'	AB	5"	4.1, 7.9
Leo	54 Leo	10h 56m	+24° 45'	AB	6"	4.5, 6.3
	83 Leo	11h 27m	+03° 01'	AB	30"	6.6, 7.5
	Tau Leo	11h 28m	+02° 51'	AB	89"	4.9, 7.5
LMi	42 LMi	10h 46m	+30° 41'	AB	197"	5.3, 7.8
Mus	HJ 4432	11h 23m	-64° 57'	AB	2"	5.4, 6.6
Sex	35 Sex	10h 43m	+04° 45'	AB	7"	6.2, 7.1
UMa	Alp UMa	11h 04m	+61° 45'	AB	381"	2.0, 7.0
	STF 1520	11h 16m	+52° 46'	AB	12"	6.5, 7.8
	Xi UMa	11h 18m	+31° 32'	AB	3"	4.3, 4.8
Vel	PZ 3	10h 32m	-45° 04'	AB	14"	5.6, 6.0
	x Vel	10h 39m	-55° 36'	AB	52"	4.4, 6.1
	Mu Vel	10h 47m	-49° 25'	AB	2"	2.8, 5.7

Open Clusters

Con.	Object	R.A.	Dec.	Diam	Mag.
Car	NGC 3293	10h 36m	-58° 14'	5'	4.7
	Col 227	10h 42m	-65° 06'	15'	8.4
	IC 2602	10h 43m	-64° 24'	50'	1.6
	Col 230	10h 44m	-59° 33'	5'	6.5
	Col 228	10h 44m	-60° 05'	14'	4.9
	Col 232	10h 45m	-59° 39'	4'	6.8
	Col 231	10h 45m	-59° 27'	14'	9.0
	Col 236	10h 57m	-61° 12'	10'	7.7
	NGC 3496	11h 00m	-60° 20'	9'	8.2
	NGC 3532	11h 06m	-58° 46'	53'	3.0
	Col 241	11h 12m	-60° 45'	5'	8.2
	Col 240	11h 12m	-60° 24'	32'	3.9
	NGC 3590	11h 13m	-60° 47'	4'	8.2
	Sto 13	11h 14m	-58° 58'	5'	7.0
	IC 2714	11h 18m	-62° 42'	12'	8.2
Cen	NGC 3680	11h 26m	-43° 15'	12'	7.6
Vel	NGC 3330	10h 39m	-54° 07'	6'	7.4

NGC 3293   NGC 3372

Open Clusters with Nebulae

Con.	Object	R.A.	Dec.	Diam	Mag.
Car	NGC 3324	10h 37m	-58° 40'	5'	6.7
	Col 233	10h 45m	-59° 42'	10'	5.0
	NGC 3572	11h 10m	-60° 15'	6'	6.6

Nebulae

Con.	Object	R.A.	Dec.	Diam	Mag.
Car	NGC 3372	10h 45m	-59° 52'	120'	3.0

M 97   NGC 3521

Planetary Nebulae

Con.	Object	R.A.	Dec.	Diam	Mag.
Car	He 3-519	10h 55m	-60° 32'	62"	10.4
	IC 2621	11h 00m	-65° 15'	5"	11.2
UMa	M 97	11h 15m	+55° 01'	180"	9.9

Galaxies

Con.	Object	R.A.	Dec.	Diam	Mag.	Type
Cen	NGC 3557	11h 10m	-37° 32'	4'	10.6	Gx E
Crt	NGC 3511	11h 03m	-23° 05'	6'	11.1	Gx S
	NGC 3513	11h 04m	-23° 15'	3'	11.6	Gx S
Hya	NGC 3585	11h 13m	-26° 45'	5'	10.2	Gx E
	NGC 3621	11h 18m	-32° 49'	12'	9.2	Gx S
Leo	NGC 3338	10h 42m	+13° 45'	6'	10.9	Gx S
	M 95	10h 44m	+11° 42'	7'	10.0	Gx S
	NGC 3367	10h 47m	+13° 45'	3'	11.6	Gx S
	M 96	10h 47m	+11° 49'	8'	9.3	Gx S
	NGC 3377	10h 48m	+13° 59'	5'	10.1	Gx E
	M 105	10h 48m	+12° 35'	5'	9.4	Gx E
	NGC 3384	10h 48m	+12° 38'	6'	9.7	Gx E
	NGC 3412	10h 51m	+13° 25'	4'	10.5	Gx S
	NGC 3489	11h 00m	+13° 54'	4'	10.4	Gx S
	NGC 3507	11h 03m	+18° 08'	3'	11.9	Gx S
	NGC 3521	11h 06m	-00° 02'	11'	9.3	Gx S
	NGC 3593	11h 15m	+12° 49'	5'	10.8	Gx S
	NGC 3596	11h 15m	+14° 47'	4'	11.0	Gx S
	NGC 3607	11h 17m	+18° 03'	5'	9.9	Gx E
	NGC 3608	11h 17m	+18° 09'	3'	10.8	Gx E
	M 65	11h 19m	+13° 06'	10'	9.3	Gx S
	NGC 3626	11h 20m	+18° 21'	3'	10.6	Gx S
	M 66	11h 20m	+12° 59'	9'	9.0	Gx S
	NGC 3628	11h 20m	+13° 35'	15'	9.8	Gx S
	NGC 3640	11h 21m	+03° 14'	4'	10.4	Gx E
	NGC 3646	11h 22m	+20° 10'	4'	11.0	Gx S
	NGC 3686	11h 28m	+17° 13'	3'	11.1	Gx S
	NGC 3705	11h 30m	+09° 17'	5'	11.1	Gx S
LMi	NGC 3344	10h 44m	+24° 55'	7'	10.2	Gx S
	NGC 3414	10h 51m	+27° 59'	4'	10.9	Gx S
	NGC 3432	10h 53m	+37° 10'	7'	11.4	Gx S
	NGC 3486	11h 00m	+28° 58'	7'	10.3	Gx S
	NGC 3504	11h 03m	+27° 58'	3'	11.0	Gx S
Sex	NGC 3423	10h 51m	+05° 50'	4'	11.0	Gx S
UMa	NGC 3310	10h 39m	+53° 30'	3'	10.4	Gx S
	NGC 3319	10h 39m	+41° 41'	6'	11.1	Gx S
	NGC 3359	10h 47m	+63° 13'	7'	10.5	Gx S
	M 108	11h 12m	+55° 40'	9'	10.2	Gx S
	NGC 3610	11h 18m	+58° 47'	3'	10.5	Gx S
	NGC 3613	11h 19m	+58° 00'	4'	10.7	Gx E
	NGC 3631	11h 21m	+53° 10'	5'	10.6	Gx S
	NGC 3642	11h 22m	+59° 04'	5'	10.9	Gx S
	NGC 3665	11h 25m	+38° 46'	3'	10.8	Gx E
	NGC 3675	11h 26m	+43° 35'	6'	10.2	Gx S

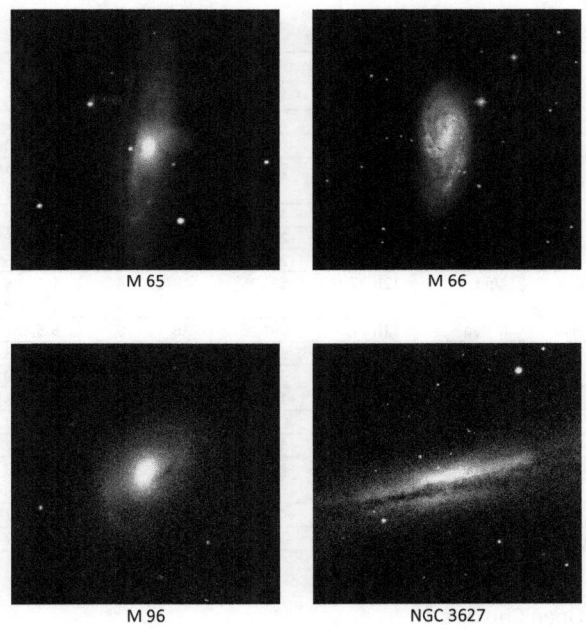

M 65

M 66

M 96

NGC 3627

List 12

Multiple Stars

Con.	Object	R.A.	Dec.	Pair	Sep.	Mag.
Cen	DUN 114	11h 40m	-38° 07'	AB	17"	6.7, 8.0
	Del Cen	12h 08m	-50° 43'	AB	269"	2.5, 4.4
				AC	216"	2.5, 6.3
	D Cen	12h 14m	-45° 43'	AB	3"	5.8, 7.0
Com	2 Com	12h 04m	+21° 28'	AB	4"	6.2, 7.5
	17 Com	12h 29m	+25° 55'	AB	146"	5.3, 6.6
Cru	Alp Cru	12h 27m	-63° 06'	AB	4"	1.3, 1.6
				AC	90"	1.3, 4.8
Hya	N Hya	11h 32m	-29° 16'	AB	9"	5.6, 5.7
	HJ 4455	11h 37m	-33° 34'	AB	3"	6.0, 7.8
	JC 17	12h 10m	-34° 42'	AB	3"	6.4, 8.0
Leo	90 Leo	11h 35m	+16° 48'	AB	3"	6.3, 7.3
Mus	HJ 4498	12h 06m	-65° 43'	AB	9"	6.1, 7.7
UMa	STF 1561	11h 39m	+45° 07'	AB	9"	6.5, 8.2
	65 UMa	11h 55m	+46° 29'	AC	4"	6.7, 8.3
				AD	63"	6.7, 7.0
	67 Uma	12h 02m	+43° 03'	AB	279"	5.2, 6.7
Vir	STF 1627	12h 18m	-03° 57'	AB	20"	6.5, 6.9

Open Clusters

Con.	Object	R.A.	Dec.	Diam	Mag.
Cen	NGC 3766	11h 36m	-61° 37'	12'	5.3
	NGC 3960	11h 51m	-55° 41'	6'	8.3
	NGC 3960	11h 51m	-55° 41'	6'	8.3
Com	Mel 111	12h 25m	+26° 06'	275'	1.8
Cru	NGC 4052	12h 02m	-63° 13'	9'	8.8
	NGC 4103	12h 07m	-61° 15'	6'	7.4
	NGC 4337	12h 24m	-58° 07'	4'	8.9
	NGC 4349	12h 24m	-61° 52'	15'	7.4
	NGC 4439	12h 28m	-60° 06'	4'	8.4
Mus	NGC 4463	12h 30m	-64° 47'	5'	7.2

Open Clusters with Nebulae

Con.	Object	R.A.	Dec.	Diam	Mag.
Cen	IC 2944	11h 37m	-63° 02'	14'	4.5
Mus	NGC 4463	12h 30m	-64° 47'	5'	7.2

Nebulae

Con.	Object	R.A.	Dec.	Diam	Mag.
Cen	IC 2944	11h 37m	-63° 02'	14'	?

Planetary Nebulae

Con.	Object	R.A.	Dec.	Diam	Mag.
Cen	NGC 3918	11h 50m	-57° 11'	13"	8.1

Globular Clusters

Con.	Object	R.A.	Dec.	Diam	Mag.
Com	NGC 4147	12h 10m	+18° 33'	4'	10.4
Mus	NGC 4372	12h 26m	-72° 40'	19'	7.2

M 106

M 98

Galaxies

Con.	Object	R.A.	Dec.	Diam	Mag.	Type
CVn	NGC 4111	12h 07m	+43° 04'	5'	10.7	Gx S
	NGC 4143	12h 10m	+42° 32'	2'	11.1	Gx S
	NGC 4145	12h 10m	+39° 53'	6'	11.2	Gx S
	NGC 4151	12h 11m	+39° 24'	6'	10.2	Gx S
	NGC 4214	12h 16m	+36° 20'	9'	9.8	Gx I
	NGC 4244	12h 17m	+37° 48'	17'	10.0	Gx S
	NGC 4242	12h 18m	+45° 37'	5'	11.1	Gx S
	M 106	12h 19m	+47° 18'	19'	8.7	Gx S
	NGC 4395	12h 26m	+33° 33'	13'	10.9	Gx S
	NGC 4449	12h 28m	+44° 06'	6'	9.8	Gx I
Cen	NGC 4373	12h 25m	-39° 46'	3'	10.9	Gx S
Com	NGC 4136	12h 09m	+29° 56'	4'	11.2	Gx S
	M 98	12h 14m	+14° 54'	10'	10.0	Gx S

Galaxies (cont.)

Con.	Object	R.A.	Dec.	Diam	Mag.	Type
Com	NGC 4203	12h 15m	+33° 12'	3'	10.7	Gx S
	NGC 4212	12h 16m	+13° 54'	3'	11.3	Gx S
	NGC 4251	12h 18m	+28° 11'	4'	10.6	Gx S
	M 99	12h 19m	+14° 25'	5'	9.7	Gx S
	NGC 4274	12h 20m	+29° 37'	7'	10.4	Gx S
	NGC 4278	12h 20m	+29° 17'	4'	10.2	Gx E
	NGC 4293	12h 21m	+18° 23'	6'	10.5	Gx S
	NGC 4314	12h 23m	+29° 54'	4'	10.5	Gx S
	M 100	12h 23m	+15° 49'	7'	9.7	Gx S
	NGC 4350	12h 24m	+16° 42'	3'	11.1	Gx S
	M 85	12h 25m	+18° 11'	7'	9.2	Gx S
	NGC 4394	12h 26m	+18° 13'	4'	10.9	Gx S
	NGC 4414	12h 26m	+31° 13'	4'	10.7	Gx S
	NGC 4419	12h 27m	+15° 03'	3'	11.4	Gx S
	NGC 4448	12h 28m	+28° 37'	4'	11.2	Gx S
	NGC 4450	12h 28m	+17° 05'	5'	10.3	Gx S
	NGC 4473	12h 30m	+13° 26'	5'	10.1	Gx E
	NGC 4477	12h 30m	+13° 38'	4'	10.4	Gx S
Crv	NGC 4027	12h 00m	-19° 16'	3'	11.2	Gx S
	NGC 4038	12h 02m	-18° 52'	5'	10.5	Gx S
	NGC 4039	12h 02m	-18° 53'	3'	10.7	Gx S
Crt	NGC 3887	11h 47m	-16° 51'	3'	10.9	Gx S
	NGC 3962	11h 55m	-13° 58'	3'	10.9	Gx E
Dra	NGC 4125	12h 08m	+65° 10'	6'	9.9	Gx E
	NGC 4236	12h 17m	+69° 27'	22'	9.9	Gx S
Hya	NGC 3904	11h 49m	-29° 17'	3'	11.0	Gx E
	NGC 3923	11h 51m	-28° 48'	6'	10.0	Gx E
	NGC 4105	12h 07m	-29° 46'	3'	10.7	Gx E
Leo	NGC 3810	11h 41m	+11° 28'	4'	10.8	Gx S
UMa	NGC 3718	11h 33m	+53° 04'	8'	10.6	Gx S
	NGC 3726	11h 33m	+47° 02'	6'	10.5	Gx S
	NGC 3729	11h 34m	+53° 08'	3'	11.6	Gx S
	NGC 3738	11h 36m	+54° 31'	3'	11.4	Gx I
	NGC 3756	11h 37m	+54° 18'	4'	11.4	Gx S
	NGC 3877	11h 46m	+47° 30'	6'	11.3	Gx S
	NGC 3893	11h 49m	+48° 43'	5'	10.4	Gx S
	NGC 3898	11h 49m	+56° 05'	4'	11.0	Gx S
	NGC 3941	11h 53m	+36° 59'	4'	10.4	Gx S
	NGC 3945	11h 53m	+60° 41'	6'	10.5	Gx S
	NGC 3949	11h 53m	+47° 52'	3'	10.9	Gx S
	NGC 3953	11h 54m	+52° 20'	7'	10.2	Gx S
	NGC 3982	11h 56m	+55° 07'	2'	11.2	Gx S
	M 109	11h 58m	+53° 23'	7'	9.9	Gx S
	NGC 3998	11h 58m	+55° 27'	3'	10.5	Gx S

Galaxies (cont.)

Con.	Object	R.A.	Dec.	Diam	Mag.	Type
UMa	NGC 4026	11h 59m	+50° 58'	5'	10.7	Gx S
	NGC 4036	12h 01m	+61° 54'	4'	10.6	Gx S
	NGC 4041	12h 02m	+62° 08'	3'	11.1	Gx S
	NGC 4051	12h 03m	+44° 32'	5'	10.7	Gx S
	NGC 4062	12h 04m	+31° 54'	4'	11.3	Gx S
	NGC 4088	12h 06m	+50° 32'	6'	11.1	Gx S
	NGC 4096	12h 06m	+47° 29'	7'	11.1	Gx S
	NGC 4100	12h 06m	+49° 35'	5'	11.0	Gx S
	NGC 4144	12h 10m	+46° 27'	6'	11.6	Gx S
Vir	NGC 4030	12h 00m	-01° 06'	4'	10.8	Gx S
	NGC 4123	12h 08m	+02° 53'	4'	11.3	Gx S
	NGC 4178	12h 13m	+10° 52'	5'	11.4	Gx S
	NGC 4179	12h 13m	+01° 18'	4'	11.2	Gx S
	NGC 4216	12h 16m	+13° 09'	8'	10.2	Gx S
	NGC 4261	12h 19m	+05° 50'	4'	10.5	Gx E
	NGC 4267	12h 20m	+12° 48'	3'	10.8	Gx S
	M 61	12h 22m	+04° 28'	7'	9.6	Gx S
	NGC 4365	12h 24m	+07° 19'	7'	9.6	Gx E
	NGC 4371	12h 25m	+11° 42'	4'	10.7	Gx S
	M 84	12h 25m	+12° 53'	7'	9.4	Gx E
	NGC 4388	12h 26m	+12° 40'	6'	11.3	Gx S
	M 86	12h 26m	+12° 57'	9'	9.0	Gx E
	NGC 4429	12h 27m	+11° 06'	6'	10.2	Gx S
	NGC 4435	12h 28m	+13° 05'	3'	10.8	Gx S
	NGC 4438	12h 28m	13° 01'	9'	10.0	Gx S
	NGC 4442	12h 28m	+09° 48'	5'	10.4	Gx S
	NGC 4457	12h 29m	+03° 34'	3'	10.8	Gx S
	NGC 4461	12h 29m	+13° 11'	4'	11.1	Gx S
	M 49	12h 30m	+08° 00'	10'	8.4	Gx E

M 100

NGC 4038 & NGC 4039

List 13

Multiple Stars

Con.	Object	R.A.	Dec.	Pair	Sep.	Mag.
Cam	32 Cam	12h 49m	+83° 25′	AB	22″	5.3, 5.7
CVn	Alp CVn	12h 56m	+38° 19′	AB	19″	2.9, 5.5
	16 CVn	13h 10m	+38° 30′	AB	278″	6.0, 6.3
Cen	J Cen	13h 23m	-60° 59′	AB	61″	4.5, 6.2
Com	24 Com	12h 35m	+18° 23′	AB	20″	5.1, 6.3
	32 Com	12h 52m	+17° 04′	AB	196″	6.3, 6.9
	Alp Com	13h 10m	+17° 32′	AB	89″	5.1, 5.1
Crv	STF 1669	12h 41m	-13° 01′	AB	5″	5.9, 5.9
Cru	Gam Cru	12h 31m	-57° 07′	AB	127″	1.6, 6.5
	Beta Cru	12h 47m	-59° 41′	AC	373″	1.3, 7.2
	Mu Cru	12h 55m	-57° 11′	AB	35″	3.9, 5.0
Dra	STFA 25	13h 14m	+67° 17′	AB	179″	6.5, 7.0
Mus	The Mus	13h 08m	-65° 18′	AB	5″	5.7, 7.6
	Eta Mus	13h 15m	-67° 54′	AC	58″	4.8, 7.2
UMa	STF 1695	12h 56m	+54° 06′	AB	3.8″	6.0, 7.8
	Zet UMa	13h 24m	+54° 56′	AB	709″	2.2, 4.0
				AC	14″	2.2, 3.9

Open Clusters

Con.	Object	R.A.	Dec.	Diam	Mag.
Cen	NGC 4852	13h 00m	-59° 37′	11′	8.9
	NGC 5138	13h 27m	-59° 02′	7′	7.6
Cru	NGC 4609	12h 42m	-63° 00′	5′	6.9
	NGC 4755	12h 54m	-60° 21′	10′	4.2

NGC 4755

NGC 5139

Globular Clusters

Con.	Object	R.A.	Dec.	Diam	Mag.
Cen	NGC 5139	13h 27m	-47° 29'	36'	3.9
Com	M 53	13h 13m	+18° 10'	13'	7.7
	NGC 5053	13h 16m	+17° 42'	11'	9.0
Hya	M 68	12h 39m	-26° 45'	12'	7.3
Mus	NGC 4833	13h 00m	-70° 53'	14'	8.4

M 53   M 68

Galaxies

Con.	Object	R.A.	Dec.	Diam	Mag.	Type
CVn	NGC 4490	12h 31m	+41° 39'	6'	9.8	Gx S
	NGC 4618	12h 42m	+41° 09'	4'	10.9	Gx S
	NGC 4631	12h 42m	+32° 32'	16'	9.1	Gx S
	NGC 4656	12h 44m	+32° 10'	15'	10.2	Gx S
	M 94	12h 51m	+41° 07'	11'	8.0	Gx S
	NGC 5005	13h 11m	+37° 04'	6'	10.3	Gx S
	NGC 5033	13h 13m	+36° 36'	11'	10.1	Gx S
	M 63	13h 16m	+42° 02'	13'	9.0	Gx S
	M 51	13h 30m	+47° 12'	11'	8.5	Gx S
	NGC 5195	13h 30m	+47° 16'	6'	10.2	Gx S
Cen	NGC 4696	12h 49m	-41° 19'	5'	10.5	Gx S
	NGC 4936	13h 04m	-30° 32'	3'	10.9	Gx E
	NGC 4945	13h 05m	-49° 28'	20'	8.9	Gx S
	NGC 4976	13h 09m	-49° 30'	6'	10.2	Gx S
	NGC 5102	13h 22m	-36° 38'	9'	9.1	Gx S
	NGC 5128	13h 25m	-43° 01'	26'	6.8	Gx E
	NGC 5161	13h 29m	-33° 10'	6'	11.6	Gx S
Com	NGC 4494	12h 31m	+25° 47'	5'	9.7	E
	M 88	12h 32m	+14° 25'	7'	9.7	Gx S
	M 91	12h 35m	+14° 30'	5'	10.4	Gx S
	NGC 4559	12h 36m	+27° 58'	11'	9.8	Gx S
	NGC 4565	12h 36m	+25° 59'	16'	9.5	Gx S

Galaxies (cont.)

Con.	Object	R.A.	Dec.	Diam	Mag.	Type
Com	NGC 4571	12h 37m	+14° 13'	4'	11.5	Gx S
	NGC 4651	12h 44m	+16° 24'	4'	10.8	Gx S
	NGC 4689	12h 48m	+13° 46'	4'	11.0	Gx S
	NGC 4710	12h 50m	+15° 10'	5'	10.9	Gx S
	NGC 4725	12h 50m	+25° 30'	11'	9.2	Gx S
	M 64	12h 57m	+21° 41'	10'	8.4	Gx S
Dra	NGC 4589	12h 37m	+74° 12'	3'	10.9	Gx E
	NGC 4570	12h 37m	+07° 15'	4'	10.9	Gx S
Hya	NGC 5061	13h 18m	-26° 50'	4'	10.5	Gx E
	NGC 5078	13h 20m	-27° 25'	4'	10.8	Gx S
	NGC 5101	13h 22m	-27° 26'	5'	10.6	Gx S
UMa	NGC 4605	12h 40m	+61° 36'	6'	10.4	Gx S
	NGC 5204	13h 30m	+58° 25'	5'	11.3	Gx S
Vir	M 87	12h 31m	+12° 23'	8'	8.8	Gx E
	NGC 4487	12h 31m	-08° 03'	4'	11.1	Gx S
	NGC 4437	12h 33m	+00° 07'	11'	10.5	Gx S
	NGC 4526	12h 34m	+07° 42'	7'	9.5	Gx S
	NGC 4527	12h 34m	+02° 39'	6'	10.8	Gx S
	NGC 4535	12h 34m	+08° 12'	7'	10.0	Gx S
	NGC 4536	12h 34m	+02° 11'	8'	10.4	Gx S
	NGC 4546	12h 35m	-03° 48'	3'	10.5	Gx E
	M 89	12h 36m	+12° 33'	5'	9.8	Gx E
	NGC 4564	12h 36m	+11° 26'	4'	11.2	Gx E
	NGC 4567	12h 37m	+11° 15'	3'	11.5	Gx S
	NGC 4568	12h 37m	+11° 14'	5'	11.2	Gx S
	M 90	12h 37m	+13° 10'	10'	9.6	Gx S
	NGC 4570	12h 37m	+07° 15'	4'	10.9	Gx S
	M 58	12h 38m	+11° 49'	6'	10.1	Gx S
	NGC 4596	12h 40m	+10° 11'	4'	10.7	Gx S
	M 104	12h 40m	-11° 37'	9'	7.9	Gx S
	M 59	12h 42m	+11° 39'	5'	9.8	Gx E
	NGC 4636	12h 43m	+02° 41'	6'	9.7	Gx E
	NGC 4643	12h 43m	+01° 59'	3'	10.8	Gx S
	M 60	12h 44m	+11° 33'	7'	8.8	Gx E
	NGC 4624	12h 45m	+03° 03'	4'	10.3	Gx S
	NGC 4666	12h 45m	-00° 28'	5'	11.2	Gx S
	NGC 4684	12h 47m	-02° 44'	3'	11.5	Gx S
	NGC 4691	12h 48m	-03° 20'	3'	10.9	Gx S
	NGC 4698	12h 48m	+08° 29'	4'	10.9	Gx S
	NGC 4697	12h 49m	-05° 48'	7'	9.2	Gx E
	NGC 4699	12h 49m	-08° 40'	4'	10.2	Gx S
	NGC 4731	12h 51m	-06° 24'	7'	11.0	Gx P
	NGC 4754	12h 52m	+11° 19'	5'	10.3	Gx S
	NGC 4753	12h 52m	-01° 12'	6'	10.5	Gx S

Galaxies (cont.)

Con.	Object	R.A.	Dec.	Diam	Mag.	Type
Vir	NGC 4762	12h 53m	+11° 14′	9′	10.0	Gx S
	NGC 4772	12h 53m	+02° 10′	3′	11.7	Gx S
	NGC 4775	12h 54m	-06° 37′	2′	11.2	Gx S
	NGC 4781	12h 54m	-10° 32′	4′	11.2	Gx S
	NGC 4818	12h 57m	-08° 32′	4′	11.2	Gx S
	NGC 4856	12h 59m	-15° 03′	4′	10.7	Gx S
	NGC 4866	12h 59m	+14° 10′	6′	10.6	Gx S
	NGC 4900	13h 01m	+02° 30′	2′	11.8	Gx S
	NGC 4902	13h 01m	-14° 31′	3′	10.9	Gx S
	NGC 4941	13h 04m	-05° 33′	4′	11.4	Gx S
	NGC 4939	13h 04m	-10° 20′	6′	11.0	Gx S
	NGC 4958	13h 06m	-08° 01′	4′	10.7	Gx S
	NGC 4995	13h 10m	-07° 50′	3′	11.3	Gx S
	NGC 5018	13h 13m	-19° 31′	3′	10.7	Gx S
	NGC 5044	13h 15m	-16° 23′	3′	10.9	Gx E
	NGC 5054	13h 17m	-16° 38′	5′	10.8	Gx S
	NGC 5068	13h 19m	-21° 02′	7′	9.8	Gx S
	NGC 5084	13h 20m	-21° 50′	9′	10.8	Gx S

NGC 4631

M 63

M 51

NGC 5128

List 14

Multiple Stars

Con.	Object	R.A.	Dec.	Pair	Sep.	Mag.
Boo	Kap Boo	14h 14m	+51° 47′	AB	14″	4.5, 6.6
	Iot Boo	14h 16m	+51° 22′	AB	40″	4.8, 7.4
	STF 1835	14h 23m	+08° 27′	AB	6″	5.0, 6.8
Cen	Q Cen	13h 42m	-54° 34′	AB	6″	5.2, 6.5
	DUN 142	13h 44m	-59°14′	AB	33″	6.5, 7.6
	3 Cen	13h 52m	-33° 00′	AB	8″	4.5, 6.0
	N Cen	13h 52m	-52° 49′	AB	18″	5.2, 7.5
	4 Cen	13h 53m	-31° 56′	AB	15″	4.8, 8.5
	DUN 159	14h 23m	-58° 28′	AB	9″	5.0, 7.6
Hya	H N 69	13h 37m	-26° 30′	AB	10″	5.7, 6.6
Lup	HJ 4672	14h 20m	-43° 04′	AB	4″	5.8, 7.9

Open Clusters

Con.	Object	R.A.	Dec.	Diam	Mag.
Cen	NGC 5281	13h 47m	-62° 55′	8′	5.9
	NGC 5316	13h 54m	-61° 52′	13′	6.0
	NGC 5460	14h 07m	-48° 21′	35′	5.6
	NGC 5606	14h 28m	-59° 38′	3′	7.7
	NGC 5617	14h 30m	-60° 43′	10′	6.3

Planetary Nebulae

Con.	Object	R.A.	Dec.	Diam	Mag.
Lup	IC 4406	14h 22m	-44° 09′	46″	10.3
Mus	NGC 5189	13h 34m	-65° 58′	140″	10.3

NGC 5466

M 3

Globular Clusters

Con.	Object	R.A.	Dec.	Diam	Mag.
Boo	NGC 5466	14h 05m	+28° 32'	11'	9.2
CVn	M 3	13h 42m	+28° 23'	18'	6.3
Cen	NGC 5286	13h 46m	-51° 22'	9'	7.4
Vir	NGC 5634	14h 30m	-05° 59'	5'	9.5

M 83   M 101

Galaxies

Con.	Object	R.A.	Dec.	Diam	Mag.	Type
Boo	**NGC 5248**	13h 38m	+08° 53'	6'	10.0	Gx S
	NGC 5557	14h 18m	+36° 30'	2'	11.2	Gx E
CVn	NGC 5353	13h 53m	+40° 17'	2'	11.2	Gx S
	NGC 5371	13h 56m	+40° 28'	4'	10.7	Gx S
Cen	IC 4296	13h 37m	-33° 58'	3'	10.6	Gx E
	NGC 5253	13h 40m	-31° 39'	5'	10.7	Gx I
	NGC 5419	14h 04m	-33° 59'	4'	10.8	Gx E
	NGC 5483	14h 10m	-43° 19'	4'	11.4	Gx S
Hya	M 83	13h 37m	-29° 52'	15'	7.8	Gx S
Lup	NGC 5530	14h 18m	-43° 23'	4'	11.3	Gx S
	IC 4402	14h 21m	-46° 18'	7'	12.0	Gx S
UMa	NGC 5204	13h 30m	+58° 25'	5'	11.3	Gx S
	NGC 5322	13h 49m	+60° 11'	6'	10.1	Gx E
	M 101	14h 03m	+54° 21'	29'	7.5	Gx S
	NGC 5474	14h 05m	+53° 40'	5'	11.0	Gx S
	NGC 5585	14h 20m	+56° 44'	6'	11.0	Gx S
Vir	NGC 5247	13h 38m	-17° 53'	6'	10.4	Gx S
	NGC 5363	13h 56m	+05° 15'	4'	10.8	Gx S
	NGC 5364	13h 56m	+05° 01'	7'	10.5	Gx S
	NGC 5427	14h 03m	-06° 02'	3'	11.5	Gx S
	NGC 5566	14h 20m	+03° 56'	7'	10.7	Gx S
	NGC 5576	14h 21m	+03° 16'	4'	10.9	Gx E

List 15

Multiple Stars

Con.	Object	R.A.	Dec.	Pair	Sep.	Mag.
Boo	Pi Boo	14h 41m	+16° 25'	AB	6"	4.9, 5.8
	Eps Boo	14h 45m	+27° 04'	AB	3"	2.6, 4.8
	STF 1884	14h 48m	+24° 22'	AB	2"	6.6, 7.5
	39 Boo	14h 50m	+48° 43'	AB	4"	6.3, 6.7
	XI Boo	14h 51m	+19° 06'	AB	6"	4.8, 7.0
	Del Boo	15h 16m	+33° 19'	AB	104"	3.6, 7.9
	Mu Boo	15h 25m	+37° 23'	AB	107"	4.3, 7.1
				BC	2"	7.1, 7.6
Cen	Alpha Cen	14h 40m	-60° 50'	AB	18"	0.0, 1.3
Cir	DUN 169	14h 45m	-55° 36'	AB	69"	6.1, 7.6
Hya	54 Hya	14h 46m	-25° 27'	AB	8"	5.1, 7.3
Lib	Alp Lib	14h 51m	-16° 03'	AB	231"	2.7, 5.2
Lup	HJ 4690	14h 37m	-46° 08'	AB	19"	5.6, 7.7
	HJ 4715	14h 57m	-47° 53'	AB	2"	6.0, 6.8
	DUN 178	15h 12m	-45°17'	AC	31"	6.5, 7.3
	DUN 177	15h 12m	-48° 44'	AB	27"	3.8, 5.5
	Zet Lup	15h 12m	-52° 06'	AB	72"	3.5, 6.7
	Mu Lup	15h 19m	-47° 53'	BC	22"	4.9, 6.3
UMi	Pi^1 UMi	15h 29m	+80° 27'	AB	31.7"	6.6, 7.3

M 5   NGC 5927

Open Clusters

Con.	Object	R.A.	Dec.	Diam	Mag.
Cen	NGC 5662	14h 36m	-56° 37'	30'	5.5
Cir	NGC 5823	15h 06m	-55° 36'	12'	7.9
Lup	NGC 5822	15h 04m	-54° 24'	39'	6.5
Nor	NGC 5925	15h 27m	-54° 32'	20'	8.4

Planetary Nebulae

Con.	Object	R.A.	Dec.	Diam	Mag.
Lup	NGC 5882	15h 17m	-45° 39'	7"	9.4

Globular Clusters

Con.	Object	R.A.	Dec.	Diam	Mag.
Hya	NGC 5694	14h 40m	-26° 32'	4'	10.2
Lib	NGC 5897	15h 17m	-21° 01'	13'	8.4
Lup	NGC 5824	15h 04m	-33° 04'	6'	9.1
	NGC 5927	15h 28m	-50° 40'	6'	8.0
Ser	M 5	15h 19m	+02° 05'	23'	5.7

Galaxies

Con.	Object	R.A.	Dec.	Diam	Mag.	Type
Boo	NGC 5676	14h 33m	+49° 27'	4'	11.1	Gx S
Dra	NGC 5678	14h 32m	+57° 55'	3'	11.5	Gx S
	M 102	15h 06m	+55° 46'	5'	10.2	Gx S
	NGC 5907	15h 16m	+56° 20'	13'	10.7	Gx S
Lup	NGC 5643	14h 33m	-44° 10'	5'	10.5	Gx S
Ser	NGC 5921	15h 22m	+05° 04'	5'	11.1	Gx S
Vir	NGC 5701	14h 39m	+05° 22'	4'	11.0	Gx S
	NGC 5713	14h 40m	-00° 17'	3'	11.1	Gx S
	NGC 5746	14h 45m	+01° 57'	7'	10.8	Gx S
	NGC 5813	15h 01m	+01° 42'	4'	10.9	Gx E
	NGC 5838	15h 05m	+02° 06'	4'	10.8	Gx E
	NGC 5846	15h 06m	+01° 36'	4'	10.3	Gx E
	NGC 5850	15h 07m	+01° 33'	4'	11.4	Gx S

M 102

NGC 5907

List 16

Del Aps   Xi Sco   Nu Sco

Multiple Stars

Con.	Object	R.A.	Dec.	Pair	Sep.	Mag.
Aps	Del Aps	16h 20m	-78° 42'	AB	103"	4.9, 5.4
CrB	Zet CrB	15h 39m	+36° 38'	AB	6"	5.0, 5.9
	Sig CrB	16h 15m	+33° 52'	AB	7"	5.6, 6.5
	Nu^1 CrB	16h 22m	+33° 48'	AB	361"	5.4, 5.6
Her	Kap Her	16h 08m	+17° 03'	AB	27"	5.1, 6.2
Lib	STF 1962	15h 39m	-08°47'	AB	12"	6.4, 6.5
Lup	HJ 4788	15h 36m	-44° 57'	AB	2"	4.7, 6.5
	HWE 79	15h 44m	-41° 49'	AB	3"	6.1, 7.9
	DUN 192	15h 47m	-35° 31'	AC	35"	6.9, 7.3
	Xi Lup	15h 57m	-33° 58'	AB	10"	5.1, 5.6
	Eta Lup	16h 00m	-38° 24'	AB	15"	3.4, 7.5
Nor	Iot Nor	16h 04m	-57° 47'	AC	11"	4.6, 8.0
Oph	Rho Oph	16h 26m	-23° 27'	AB	3"	5.1, 5.7
				AC	150"	5.1, 7.3
Sco	2 Sco	15h 54m	-25° 20'	AB	2"	4.7, 7.0
	Xi Sco	16h 04m	-11° 22'	AC	8"	4.9, 7.3
	Bet Sco	16h 05m	-19° 48'	AC	14"	2.6, 4.5
	Nu Sco	16h 12m	-19° 28'	AC	41"	4.2, 6.6
				CD	2"	6.6, 7.2
	BSO 12	16h 20m	-30° 54'	AB	24"	5.6, 6.9
	H N 39	16h 25m	-29° 42'	AB	4"	5.9, 6.6
	Alp Sco	16h 29m	-26° 26'	AB	3"	1.0, 5.4
Ser	Del Ser	15h 35m	+10° 32'	AB	4.0"	4.2, 5.2

NGC 6067

NGC 6134

Open Clusters

Con.	Object	R.A.	Dec.	Diam	Mag.
Nor	Col 292	15h 51m	-57° 40'	15'	7.9
	NGC 6067	16h 13m	-54° 13'	12'	5.6
	NGC 6087	16h 19m	-57° 56'	12'	5.4
	NGC 6134	16h 28m	-49° 09'	6'	7.2
Sco	NGC 6124	16h 25m	-40° 39'	40'	5.8
	Col 302	16h 26m	-26° 00'	505'	1.0
TrA	NGC 6025	16h 03m	-60° 26'	12'	5.1

Planetary Nebulae

Con.	Object	R.A.	Dec.	Diam	Mag.
Her	IC 4593	16h 12m	+12° 04'	11"	10.7

NGC 6101

NGC 5986

Globular Clusters

Con.	Object	R.A.	Dec.	Diam	Mag.
Aps	NGC 6101	16h 26m	-72° 12'	11'	9.2
Lup	NGC 5986	15h 46m	-37° 47'	10'	7.6
Nor	NGC 5946	15h 35m	-50° 40'	3'	8.4
Sco	M 80	16h 17m	-22° 59'	9'	7.3
	M 4	16h 24m	-26° 32'	26'	5.4
	NGC 6144	16h 27m	-26° 01'	9'	9.0
	NGC 6139	16h 28m	-38° 51'	6'	9.1

M 80   M 4

Galaxies

Con.	Object	R.A.	Dec.	Diam	Mag.	Type
Dra	NGC 5982	15h 39m	+59° 21'	3'	11.3	Gx E
	NGC 5985	15h 40m	+59° 20'	6'	11.1	Gx S
	NGC 6015	15h 51m	+62° 19'	5'	11.2	Gx S
Ser	NGC 5970	15h 38m	+12° 11'	3'	11.6	Gx S

List 17

DUN 216   16 & 17 Dra   36 & 37 Dra

Multiple Stars

Con.	Object	R.A.	Dec.	Pair	Sep.	Mag.
Ara	DUN 206	16h 41m	-48° 46'	AC	10"	5.7, 6.8
	DUN 216	17h 27m	-45° 51'	AC	102"	5.6, 7.1
Dra	16 Dra	16h 36m	+52° 55'	AB	90"	5.4, 5.5
				AC	3"	5.4, 6.4
	Mu Dra	17h 05m	+54° 28'	AB	2"	5.7, 5.7
Her	36 Her	16h 41m	+04° 13'	AB	70"	5.8, 6.9
	STFA 33	17h 04m	+13° 36'	AB	305"	5.9, 6.2
	Alp Her	17h 15m	+14° 23'	AB	5"	3.5, 5.4
	Rho Her	17h 24m	+37° 09'	AB	4"	4.5, 5.4
Oph	36 Oph	17h 15m	-26° 36'	AB	5"	5.1, 5.1
				AC	733"	5.1, 6.5
	Omi Oph	17h 18m	-24° 17'	AB	10"	5.2, 6.6
Sco	CPO 70	16h 44m	-41° 07'	AB	97.2"	6.1, 6.2
	HJ 4889	16h 51m	-37° 31'	AB	6.8"	6.2, 7.8
	WFC 183	16h 54m	-41° 51'	AB	30.5"	6.4, 7.7
Tel	DUN 216	17h 27m	-45° 51'	AB	2"	5.8, 6.4
				AC	103"	5.8, 7.6

NGC 6281   M 13

Open Clusters

Con.	Object	R.A.	Dec.	Diam	Mag.
Ara	NGC 6193	16h 41m	-48° 46'	14'	5.2
	NGC 6200	16h 44m	-47° 28'	12'	7.4
	NGC 6204	16h 46m	-47° 01'	5'	8.2
	NGC 6208	16h 49m	-53° 44'	15'	7.2
	NGC 6250	16h 58m	-45° 56'	7'	5.9
	IC 4651	17h 24m	-49° 57'	12'	8.0
Nor	NGC 6152	16h 33m	-52° 39'	29'	8.1
	NGC 6169	16h 34m	-44° 02'	6'	6.6
	NGC 6167	16h 35m	-49° 46'	7'	6.7
Sco	NGC 6178	16h 36m	-45° 39'	4'	7.2
	NGC 6192	16h 40m	-43° 22'	7'	8.5
	NGC 6231	16h 54m	-41° 49'	14'	2.6
	Col 316	16h 55m	-40° 50'	105'	3.4
	NGC 6242	16h 56m	-39° 28'	9'	6.4
	NGC 6249	16h 58m	-44° 49'	6'	8.2
	NGC 6259	17h 01m	-44° 39'	10'	8.0
	NGC 6281	17h 05m	-37° 59'	8'	5.4
	NGC 6322	17h 18m	-42° 56'	10'	6.0

Open Clusters with Nebulae

Con.	Object	R.A.	Dec.	Diam	Mag.
Sco	Col 318	16h 57m	-40° 40'	60'	8.6

Planetary Nebulae

Con.	Object	R.A.	Dec.	Diam	Mag.
Her	NGC 6369	17h 29m	-23° 46'	58"	11.4

M 62

M 92

Globular Clusters

Con.	Object	R.A.	Dec.	Diam	Mag.
Ara	NGC 6352	17h 25m	-48° 25'	7'	7.8
Her	M 13	16h 42m	+36° 28'	17'	5.8
	NGC 6229	16h 47m	+47° 32'	5'	9.4
	M 92	17h 17m	+43° 08'	14'	6.5
Oph	M 107	16h 33m	-13° 03'	10'	7.8
	M 12	16h 47m	-01° 57'	15'	6.1
	NGC 6235	16h 53m	-22° 11'	5'	8.9
	M 10	16h 57m	-04° 06'	15'	6.6
	M 62	17h 01m	-30° 07'	14'	6.4
	M 19	17h 03m	-26° 16'	14'	6.8
	NGC 6284	17h 04m	-24° 46'	6'	8.9
	NGC 6287	17h 05m	-22° 42'	5'	9.3
	NGC 6293	17h 10m	-26° 35'	8'	8.3
	NGC 6304	17h 15m	-29° 28'	7'	8.3
	NGC 6316	17h 17m	-28° 08'	5'	8.1
	NGC 6325	17h 18m	-23° 46'	4'	10.2
	M 9	17h 19m	-18° 31'	9'	7.8
	NGC 6342	17h 21m	-19° 35'	4'	9.5
	NGC 6356	17h 24m	-17° 49'	7'	8.2
	NGC 6355	17h 24m	-26° 21'	5'	8.6
	NGC 6366	17h 28m	-05° 05'	8'	9.5

M 107

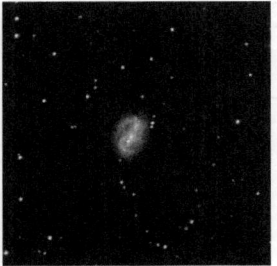
NGC 6217

Galaxies

Con.	Object	R.A.	Dec.	Diam	Mag.	Type
Ara	NGC 6215	16h 51m	-59° 00'	2'	11.2	Gx S
	NGC 6221	16h 53m	-59° 13'	4'	10.6	Gx S
	NGC 6300	17h 17m	-62° 49'	5'	10.3	Gx S
Dra	NGC 6340	17h 10m	+72° 18'	3'	11.2	Gx S
Her	NGC 6207	16h 43m	+36° 50'	3'	11.7	Gx S
UMi	NGC 6217	16h 33m	+78° 12'	3'	11.5	Gx S

List 18

Nu Dra

DUN 219

Multiple Stars

Con.	Object	R.A.	Dec.	Pair	Sep.	Mag.
Dra	Nu Dra	17h 32m	+55° 11'	AB	63"	4.9, 4.9
	Psi¹ Dra	17h 42m	+72° 09'	AB	30"	4.6, 5.6
	40 Dra	18h 00m	+80° 00'	AB	20"	5.7, 6.0
	39 Dra	18h 24m	+58° 48'	AC	90"	5.1, 8.0
Her	STT 157	17h 41m	+31° 17'	AB	117"	6.4, 7.9
	95 Her	18h 02m	+21° 36'	AB	6"	4.9, 5.2
	100 Her	18h 08m	+26° 06'	AB	14"	5.8, 5.8
Oph	53 Oph	17h 35m	+09° 35'	AB	42"	5.8, 7.5
	SHJ 251	17h 39m	+02° 02'	AB	111"	6.3, 7.7
	61 Oph	17h 45m	+02° 35'	AB	21"	6.1, 6.5
	70 Oph	18h 06m	+02° 30'	AB	5"	4.2, 6.2
Sgr	DUN 219	17h 59m	-36° 52'	AB	53"	5.8, 7.7
	PZ 6	17h 59m	-30° 15'	AB	6"	5.4, 7.0
	Eta Sgr	18h 18m	-36° 46'	AB	4"	3.1, 7.8
	WNO 6	18h 29m	-26° 35'	AB	42"	6.7, 8.0
Ser	59 Ser	18h 27m	+00° 12'	AB	3.9"	5.4, 7.6

M 23

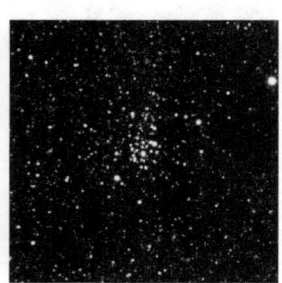

NGC 6520

Open Clusters

Con.	Object	R.A.	Dec.	Diam	Mag.
Oph	IC 4665	17h 46m	+05° 43'	70'	4.2
	Col 350	17h 48m	+01° 18'	45'	6.1
	Mel 186	18h 01m	+03° 00'	240'	3.0
	NGC 6633	18h 27m	+06° 31'	20'	4.6
Sgr	NGC 6469	17h 53m	-22° 17'	8'	8.2
	M 23	17h 57m	-18° 59'	30'	5.5
	NGC 6520	18h 03m	-27° 53'	6'	7.6
	M 21	18h 04m	-22° 29'	15'	5.9
	NGC 6546	18h 07m	-23° 18'	15'	8.0
	NGC 6568	18h 13m	-21° 35'	12'	8.6
	M 24	18h 18m	-18° 24'	5'	4.6
	Col 378	18h 25m	-19° 43'	6'	7.8
Sco	Col 332	17h 31m	-37° 05'	10'	8.9
	NGC 6383	17h 35m	-32° 35'	3'	5.5
	Col 336	17h 36m	-33° 31'	6'	6.7
	Col 338	17h 38m	-37° 34'	25'	8.0
	NGC 6400	17h 40m	-36° 57'	12'	8.8
	M 6	17h 40m	-32° 15'	33'	4.2
	Col 343	17h 42m	-40° 09'	12'	7.5
	NGC 6416	17h 44m	-32° 22'	15'	5.7
	NGC 6425	17h 47m	-31° 32'	15'	7.2
	M 7	17h 54m	-34° 48'	80'	3.3
	Col 355	17h 57m	-35° 15'	7'	8.8
Ser	NGC 6605	18h 18m	-15° 01'	29'	6.0

Open Clusters with Nebulae

Con.	Object	R.A.	Dec.	Diam	Mag.
Sgr	NGC 6530	18h 05m	-24° 21'	14'	4.6
	Col 367	18h 09m	-23° 59'	37'	6.4
	M 8	18h 03m	-24° 23'	45'	5.0
	NGC 6595	18h 17m	-19° 52'	4'	7.0
	M 18	18h 20m	-17° 06'	9'	6.9
	M 17	18h 21m	-16° 10'	18'	6.0
Sco	Col 337	17h 37m	-32° 29'	6'	7.7
	NGC 6451	17h 51m	-30° 13'	7'	8.2
Sct	NGC 6625	18h 23m	-12° 01'	39'	9.0
Ser	NGC 6604	18h 18m	-12° 15'	4'	6.5
	M 16	18h 19m	-13° 49'	55'	6.0

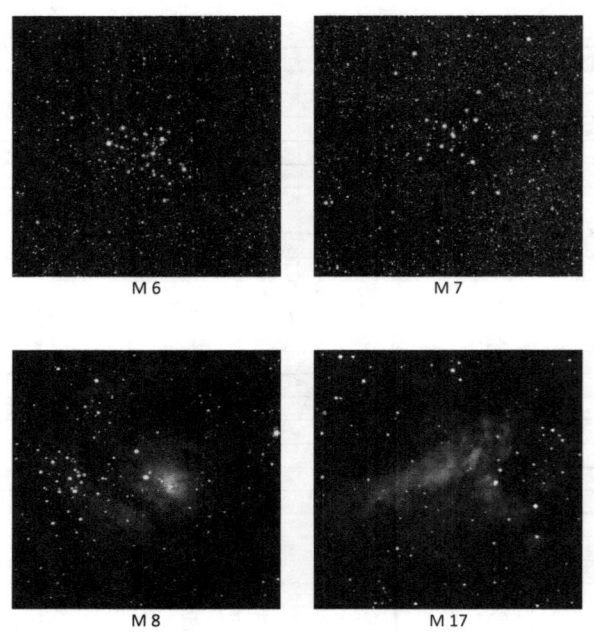

M 6

M 7

M 8

M 17

Nebulae

Con.	Object	R.A.	Dec.	Diam	Mag.
Sgr	M 20	18h 02m	-23° 02'	20'	6.3

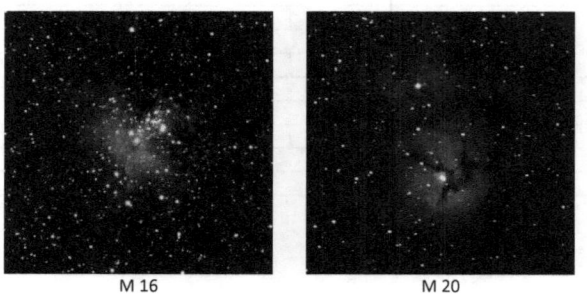

M 16

M 20

Planetary Nebulae

Con.	Object	R.A.	Dec.	Diam	Mag.
Dra	NGC 6543	17h 59m	+66° 38'	20"	8.1
Oph	He 2-260	17h 39m	-18° 18'	10"	11.0
	NGC 6572	18h 12m	+06° 51'	15"	8.1
Sgr	NGC 6629	18h 26m	-23° 12'	15"	11.3

NGC 6362   NGC 6397

Globular Clusters

Con.	Object	R.A.	Dec.	Diam	Mag.
Ara	NGC 6362	17h 32m	-67° 03'	11'	8.1
	NGC 6397	17h 41m	-53° 40'	26'	5.3
CrA	NGC 6541	18h 08m	-43° 43'	13'	6.3
Oph	M 14	17h 38m	-03° 15'	12'	7.6
	NGC 6401	17h 39m	-23° 55'	6'	7.4
	NGC 6517	18h 02m	-08° 58'	4'	10.1
Sgr	NGC 6440	17h 49m	-20° 22'	5'	9.3
	NGC 6522	18h 04m	-30° 02'	6'	9.9
	NGC 6528	18h 05m	-30° 03'	4'	9.6
	NGC 6544	18h 07m	-25° 00'	9'	7.5
	NGC 6553	18h 09m	-25° 54'	9'	8.3
	NGC 6558	18h 10m	-31° 46'	4'	8.6
	NGC 6569	18h 14m	-31° 50'	6'	8.4
	NGC 6624	18h 24m	-30° 22'	9'	7.6
	M 28	18h 25m	-24° 52'	11'	6.9
Sco	NGC 6388	17h 36m	-44° 44'	9'	6.8
	NGC 6441	17h 50m	-37° 03'	8'	7.2
	NGC 6453	17h 51m	-34° 36'	4'	10.2
	NGC 6496	17h 59m	-44° 16'	7'	8.6
Ser	NGC 6535	18h 04m	-00° 18'	4'	9.3
	NGC 6539	18h 05m	-07° 35'	7'	8.9
Tel	NGC 6584	18h 19m	-52° 13'	8'	7.9

NGC 6541

M 14

Galaxies

Con.	Object	R.A.	Dec.	Diam	Mag.	Type
Dra	NGC 6503	17h 49m	+70° 09′	6′	10.6	Gx S
	NGC 6643	18h 20m	+74° 34′	4′	11.5	Gx S
Oph	NGC 6384	17h 32m	+07° 04′	6′	11.1	Gx S
Pav	IC 4662	17h 47m	-64° 38′	3′	11.9	Gx I

List 19

Multiple Stars

Con.	Object	R.A.	Dec.	Pair	Sep.	Mag.
Aql	5 Aql	18h 47m	-00° 58'	AB	13"	5.9, 7.0
	15 Aql	19h 05m	-04° 02'	AB	39"	5.5, 7.0
	STT 178	19h 15m	+15°05'	AB	90"	5.7, 7.6
	24 Aql	19h 19m	+00° 20'	AB	427"	6.5, 6.8
CrA	Kappa CrA	18h 33m	-38° 44'	AB	21"	5.6, 6.3
	BSO 14	19h 01m	-37° 04'	AB	13"	6.3, 6.6
Cyg	STF 2486	19h 12m	+49° 51'	AB	8"	6.5, 6.7
Lyr	STF 2372	18h 42m	+34° 45'	AB	25"	6.5, 8.0
	Eps Lyr	18h 44m	+39° 40'	AB	211"	5.0, 5.3
				AC	2"	5.0, 6.1
				BD	2"	5.3, 5.4
	Zet Lyr	18h 45m	+37° 36'	AD	434"	4.3, 5.6
	Bet Lyr	18h 50m	+33° 22'	AB	46"	3.6, 6.7
	STT 525	18h 55m	+33° 58'	AC	45"	6.1, 7.6
Sge	2 & 3 Sge	19h 24m	+16° 56'	AB	342"	6.3, 6.9
Sgr	S 710	19h 07m	-16° 14'	AB	6"	6.1, 8.4
	Beta Sgr	19h 23m	-44° 28'	AB	29"	4.0, 7.2
	H N 119	19h 30m	-26° 59'	AB	7"	5.6, 8.8
Sct	H VI 50	18h 50m	-05° 55'	AB	111"	6.2, 8.2
Ser	STF 2375	18h 46m	+05° 30'	AB	3"	6.3, 6.7
	The Ser	18h 56m	+04° 12'	AB	22"	4.6, 4.9
Tel	HJ 5114	19h 28m	-54° 20'	AB	75"	5.9, 8.2
Vul	Alp Vul	19h 29m	+24° 40'	AB	425"	4.6, 5.9

NGC 6791   M 11

Open Clusters

Con.	Object	R.A.	Dec.	Diam	Mag.
Aql	NGC 6709	18h 51m	+10° 19'	13'	6.7
	NGC 6738	19h 01m	+11° 37'	15'	8.3
	NGC 6755	19h 08m	+04° 16'	14'	7.5
Lyr	Ste. 1	18h 54m	+36° 52'	20'	3.8
	NGC 6791	19h 21m	+37° 46'	10'	9.5
Sgr	NGC 6647	18h 33m	-17° 14'	9'	8.0
	M 25	18h 32m	-19° 15'	32'	6.2
	NGC 6645	18h 33m	-16° 53'	15'	8.5
	Col 394	18h 53m	-20° 23'	22'	6.3
	NGC 6716	18h 55m	-19° 54'	6'	7.5
Sct	NGC 6649	18h 33m	-10° 24'	5'	8.9
	NGC 6664	18h 37m	-08° 13'	16'	7.8
	M 26	18h 45m	-09° 23'	14'	8.0
	M 11	18h 51m	-06° 16'	13'	5.8
Ser	IC 4756	18h 39m	+05° 27'	40'	4.6
Vul	Col 399	19h 25m	+20° 11'	60'	3.6

Planetary Nebulae

Con.	Object	R.A.	Dec.	Diam	Mag.
Lyr	M 57	18h 54m	+33° 02'	86"	8.8

M 57   M 56

Globular Clusters

Con.	Object	R.A.	Dec.	Diam	Mag.
Aql	NGC 6760	19h 11m	+01° 02'	7'	9.0
Lyr	M 56	19h 17m	+30°11'	7'	8.4
Pav	NGC 6752	19h 11m	-59° 59'	20'	5.3
Sgr	NGC 6638	18h 31m	-25° 30'	7'	9.2

Globular Clusters (cont.)

Con.	Object	R.A.	Dec.	Diam	Mag.
Sgr	M 69	18h 31m	-32° 21'	7'	7.6
	NGC 6642	18h 32m	-23° 29'	5'	8.9
	NGC 6652	18h 36m	-32° 59'	4'	8.5
	M 22	18h 36m	-23° 54'	24'	5.2
	M 70	18h 43m	-32° 18'	8'	7.8
	NGC 6717	18h 55m	-22° 42'	4'	8.4
	M 54	18h 55m	-30° 29'	9'	7.7
	NGC 6723	19h 00m	-36° 38'	11'	6.8
Sct	NGC 6712	18h 53m	-08° 42'	7'	8.1

NGC 6752   M 22

Galaxies

Con.	Object	R.A.	Dec.	Diam	Mag.	Type
Pav	NGC 6684	18h 49m	-65° 10'	4'	10.5	Gx S
	NGC 6744	19h 10m	-63° 51'	20'	8.8	Gx S
	NGC 6753	19h 11m	-57° 03'	3'	11.2	Gx S

List 20

| Bet Cyg | 16 Cyg | 54 Sgr |

Multiple Stars

Con.	Object	R.A.	Dec.	Pair	Sep.	Mag.
Aql	57 Aql	19h 55m	-08° 14'	AB	36"	5.7, 6.3
Cap	Alp Cap	20h 18m	-12° 33'	AB	382"	3.7, 4.4
	Bet Cap	20h 21m	-14° 47'	AB	207"	3.2, 6.1
	Rho Cap	20h 29m	-17° 49'	AD	259"	5.0, 6.7
	Omi Cap	20h 30m	-18° 35'	AB	19"	5.9, 6.7
Cyg	Bet Cyg	19h 31m	+27° 58'	AB	35"	3.4, 4.7
	16 Cyg	19h 42m	+50° 32'	AB	39"	6.0, 6.2
	Del Cyg	19h 45m	+45° 08'	AB	3"	2.9, 6.3
	STF 2578	19h 46m	+36° 05'	AB	15"	6.4, 7.0
	Psi Cyg	19h 56m	+52° 26'	AB	2.9"	5.0, 7.5
	Omi¹ Cyg	20h 13m	+46° 44'	AC	107"	3.8, 7.0
	29 Cyg	20h 15m	+36°48'	AB	216"	5.0, 6.7
	STF 2671	20h 18m	+55° 24'	AB	4"	6.0, 7.5
	Gam Cyg	20h 22m	+40° 15'	AB	141"	2.2, 5.4
	STT 207	20h 23m	+42° 59'	AB	87"	6.4, 8.0
Dra	Eps Dra	19h 48m	+70° 16'	AB	3"	4.0, 6.9
	75 Dra	20h 28m	+81° 25'	AB	197"	5.5, 6.7
Sge	15 Sge	20h 04m	+17° 04'	AC	214"	5.9, 6.9
	The Sge	20h 10m	+20° 54'	AC	90"	6.6, 7.5
	Tha Sge	20h 10m	+20° 55'	AB	12"	6.6, 8.9
				AC	89"	6.6, 7.5
Sgr	54 Sgr	19h 41m	-16° 18'	AC	45"	5.4, 7.7
	HJ 5188	20h 21m	-29° 12'	AC	27"	6.7, 7.6
Tel	DUN 227	19h 53m	-54° 58'	AB	23"	5.8, 6.4

Open Clusters

Con.	Object	R.A.	Dec.	Diam	Mag.
Aql	Col 401	19h 38m	+00° 21'	2'	7.0
Cyg	NGC 6811	19h 37m	+46° 23'	12'	6.8
	NGC 6819	19h 41m	+40° 11'	5'	7.3
	NGC 6834	19h 52m	+29° 24'	5'	7.8
	NGC 6866	20h 04m	+44° 10'	6'	7.6
	Col 419	20h 18m	+40° 44'	5'	5.4
Oct	Col 411	20h 17m	-79° 02'	50'	5.3
Sge	Col 408	19h 53m	+18° 19'	6'	7.7
Vul	NGC 6830	19h 51m	+23° 06'	12'	7.9
	NGC 6885	20h 12m	+26° 29'	20'	8.1

Open Clusters with Nebulae

Con.	Object	R.A.	Dec.	Diam	Mag.
Cyg	NGC 6871	20h 06m	+35° 47'	30'	5.2
	NGC 6883	20h 11m	+35° 50'	35'	8.0
	IC 4996	20h 17m	+37° 38'	5'	7.3
	Ber 86	20h 20m	+38° 41'	7'	7.9
	NGC 6910	20h 23m	+40° 47'	7'	7.4
	M 29	20h 24m	+38° 30'	10'	6.6
Vul	NGC 6823	19h 43m	+23° 18'	12'	7.1

NGC 6823

NGC 6888

Nebulae

Con.	Object	R.A.	Dec.	Diam	Mag.
Cyg	NGC 6888	20h 12m	+38° 21'	20'	?

Planetary Nebulae

Con.	Object	R.A.	Dec.	Diam	Mag.
Aql	NGC 6803	19h 31m	+10° 03'	5"	11.4
Cyg	Min 1-92	19h 36m	+29° 33'	8"	6.3
	NGC 6826	19h 45m	+50° 32'	26"	8.8
Del	NGC 6891	20h 15m	+12° 42'	68"	10.5
Sge	NGC 6879	20h 10m	+16° 55'	5"	12.5
Sgr	NGC 6818	19h 44m	-14° 09'	22"	9.3
Vul	M 27	20h 00m	+22° 43'	480"	7.4

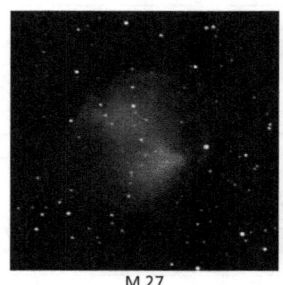 

M 27   M 55

Globular Clusters

Con.	Object	R.A.	Dec.	Diam	Mag.
Sge	M 71	19h 54m	+18° 47'	7'	8.4
Sgr	M 55	19h 40m	-30° 58'	19'	6.3
	M 75	20h 06m	-21° 55'	6'	8.6

Galaxies

Con.	Object	R.A.	Dec.	Diam	Mag.	Type
Cap	NGC 6907	20h 25m	-24° 49'	3'	11.3	Gx S
Pav	IC 4901	19h 54m	-58° 43'	6'	12.0	Gx S
	NGC 6876	20h 18m	-70° 52'	2'	11.2	Gx S
Sgr	NGC 6822	19h 45m	-14° 49'	16'	9.9	Gx I
	NGC 6902	20h 24m	-43° 39'	6'	11.5	Gx S
Tel	NGC 6868	20h 10m	-48° 23'	4'	10.8	Gx E

List 21

| 61 Cyg | DUN 236 | Eta Mic |

Multiple Stars

Con.	Object	R.A.	Dec.	Pair	Sep.	Mag.
Aqr	12 Aqr	21h 04m	-05° 49'	AB	2"	5.5, 7.3
Cyg	48 Cyg	20h 38m	+31° 34'	AB	183"	6.3, 6.5
	STF 2741	20h 59m	+50° 28'	AB	2"	5.9, 6.8
	61 Cyg	21h 07m	+38° 45'	AB	31"	5.3, 6.1
	STF 2762	21h 09m	+30° 12'	AB	4"	5.8, 7.8
Del	Gam Del	20h 47m	+16° 07'	AB	9"	4.4, 5.0
	STF 2735	20h 55m	+03° 32'	AB	2"	6.2, 8.0
Equ	Eps Equ	20h 59m	+04° 18'	AC	11"	5.3, 7.1
	Gam Equ	21h 10m	+10° 08'	AD	338"	4.7, 6.0
Ind	The Ind	21h 20m	-53° 27'	AB	7"	4.5, 6.9
Mic	DUN 236	21h 02m	-43° 00'	AB	58"	6.7, 7.0
	Eta Mic	21h 06m	-41° 23'	AB	128"	5.7, 7.9
Oct	Mu² Oct	20h 42m	-75° 21'	AB	17"	6.5, 7.1
Pav	Rmk 26	20h 52m	-62° 26'	AB	2"	6.2, 6.6
Peg	1 Peg	21h 22m	+19° 48'	AB	36"	4.2, 7.6

Open Clusters

Con.	Object	R.A.	Dec.	Diam	Mag.
Cyg	NGC 7039	21h 11m	+45° 37'	25'	7.6
	NGC 7062	21h 23m	+46° 23'	6'	8.3
	NGC 7063	21h 24m	+36° 29'	7'	7.0
	NGC 7082	21h 29m	+47° 08'	25'	7.2
Vul	NGC 6940	20h 34m	+28° 17'	31'	6.3

Open Clusters with Nebulae

Con.	Object	R.A.	Dec.	Diam	Mag.
Cyg	Col 428	21h 03m	+44° 36'	10'	8.7

Nebulae

Con.	Object	R.A.	Dec.	Diam	Mag.
Cyg	IC 5070	20h 51m	+44° 21'	60'	?
	NGC 7000	21h 02m	+44° 12'	100'	4.0

IC 5070   NGC 7000

Planetary Nebulae

Con.	Object	R.A.	Dec.	Diam	Mag.
Aqr	NGC 7009	21h 04m	-11° 22'	34"	8.0
Cyg	NGC 7027	21h 07m	+42° 14'	14"	8.5

Supernova Remnants

Con.	Object	R.A.	Dec.	Diam	Mag.
Cyg	NGC 6992	20h 56m	+31° 45'	60'	7.0

NGC 6992   NGC 6934

Globular Clusters

Con.	Object	R.A.	Dec.	Diam	Mag.
Aqr	M 72	20h 53m	-12° 32'	6'	9.2
Del	NGC 6934	20h 34m	+07° 24'	7'	8.9
	NGC 7006	21h 01m	+16° 11'	3'	10.6
Peg	M 15	21h 30m	+12° 10'	12'	6.3

Galaxies

Con.	Object	R.A.	Dec.	Diam	Mag.	Type
Cep	NGC 6951	20h 37m	+66° 06'	4'	11.3	Gx S
Cyg	NGC 6946	20h 35m	+60° 09'	12'	9.1	Gx S
Ind	NGC 7049	21h 19m	-48° 34'	4'	10.6	Gx S
Mic	NGC 6925	20h 34m	-31° 59'	5'	11.5	Gx S
Pav	IC 5052	20h 52m	-69° 12'	7'	10.6	Gx I

M 15

NGC 6946

List 22

Multiple Stars

Con.	Object	R.A.	Dec.	Pair	Sep.	Mag.
Aqr	41 Aqr	22h 14m	-21° 04'	AB	5"	5.6, 6.7
	Zet Aqr	22h 29m	-00° 01'	AB	2"	4.3, 4.5
Cep	STF 2816	21h 39m	+57° 29'	AC	12"	5.7, 7.5
				AD	20"	5.7, 7.5
	STF 2840	21h 52m	+55° 48'	AB	18"	5.6, 6.4
	Xi Cep	22h 04m	+64° 38'	AB	8"	4.5, 6.4
	STF 2893	22h 13m	+73° 18'	AB	30"	6.2, 7.9
	Del Cep	22h 29m	+58° 25'	AC	41"	4.2, 6.1
Gru	HJ 5319	22h 12m	-38° 18'	AB	2"	6.9, 7.8
Oct	Lam Oct	21h 51m	-82° 43'	AB	3"	5.6, 7.3
Peg	3 Peg	21h 38m	+06° 37'	AB	39"	6.2, 7.6
	STF 2841	21h 54m	+19° 43'	AB	23"	6.5, 8.0
	Pi Peg	22h 09m	+33° 10'	AB	573"	4.3, 5.7
	33 Peg	22h 24m	+20° 51'	AC	90"	6.3, 8.5
PsA	Eta PsA	22h 01m	-28° 27'	AB	2"	5.7, 6.8
Tuc	Del Tuc	22h 27m	-64° 58'	AB	7"	4.5, 8.7

Open Clusters

Con.	Object	R.A.	Dec.	Diam	Mag.
Cep	NGC 7160	21h 54m	+62° 36'	7'	6.1
	NGC 7235	22h 12m	+57° 16'	4'	7.7
	NGC 7261	22h 20m	+58° 07'	5'	8.4
Cyg	M 39	21h 32m	+48° 27'	31'	4.6
Lac	NGC 7209	22h 05m	+46° 29'	25'	7.7
	IC 1434	22h 11m	+52° 50'	7'	9
	NGC 7243	22h 15m	+49° 54'	21'	6.4

M 39

IC 1434

Open Clusters with Nebulae

Con.	Object	R.A.	Dec.	Diam	Mag.
Cep	IC 1396	21h 39m	+57° 30′	50′	?
	Ber 94	22h 23m	+55° 52′	2′	8.7
Cyg	IC 5146	21h 53m	+47° 16′	9′	?

Planetary Nebulae

Con.	Object	R.A.	Dec.	Diam	Mag.
Aqr	NGC 7293	22h 30m	-20° 50′	1080′	7.3
Gru	IC 5148	22h 00m	-39° 23′	120″	11.0

NGC 7293   M 30

Globular Clusters

Con.	Object	R.A.	Dec.	Diam	Mag.
Aqr	M 2	21h 33m	-00° 49′	13′	6.6
Cap	M 30	21h 40m	-23° 11′	11′	6.9

Galaxies

Con.	Object	R.A.	Dec.	Diam	Mag.	Type
Aqr	NGC 7184	22h 03m	-20° 48′	6′	11.4	Gx S
Gru	NGC 7144	21h 53m	-48° 15′	4′	10.8	Gx S
	NGC 7213	22h 09m	-47° 10′	3′	10.8	Gx S
	IC 5201	22h 21m	-46° 04′	11′	10.8	Gx S
Ind	NGC 7083	21h 36m	-63° 54′	4′	11.4	Gx S
	NGC 7090	21h 36m	-54° 33′	7′	11.0	Gx S
	IC 5152	22h 03m	-51° 17′	6′	10.5	Gx I
	NGC 7205	22h 09m	-57° 27′	4′	11.1	Gx S
Peg	NGC 7177	22h 01m	+17° 44′	3′	11.4	Gx S
	NGC 7217	22h 08m	+31° 22′	4′	10.5	Gx S

List 23

Theta Gru

DUN 241

Multiple Stars

Con.	Object	R.A.	Dec.	Pair	Sep.	Mag.
Aqr	83 Aqr	23h 05m	-07° 42'	AC	262"	5.4, 7.6
	94 Aqr	23h 19m	-13° 28'	AB	12"	5.3, 7.0
Cas	AR Cas	23h 30m	+58° 33'	AC	76"	4.9, 7.2
Cep	STF 2950	22h 51m	+61° 42'	AB	2"	6.1, 7.4
	Omi Cep	23h 19m	+68° 07'	AB	2"	4.8, 6.8
Gru	The Gru	23h 07m	-43° 31'	AC	160"	4.5, 7.8
	DUN 246	23h 07m	-50° 41'	AB	9"	6.3, 7.1
	Rst 5560	23h 21m	-50° 18'	AC	17"	6.1, 6.6
	DUN 249	23h 24m	-53° 49'	AB	27"	6.1, 7.1
Lac	8 Lac	22h 36m	+39° 38'	AB	23"	5.7, 6.3
Peg	STF 2978	23h 08m	+32° 50'	AB	9"	6.4, 7.5
PsA	Bet PsA	22h 32m	-32° 21'	AB	30"	4.3, 7.1
	DUN 241	22h 37m	-31° 40'	AB	93"	5.9, 7.6

Open Clusters

Con.	Object	R.A.	Dec.	Diam	Mag.
And	NGC 7686	23h 30m	+49° 08'	14'	5.6
Cas	M 52	23h 25m	+61° 36'	16'	6.9

Open Clusters with Nebulae

Con.	Object	R.A.	Dec.	Diam	Mag.
Cep	NGC 7380	22h 47m	+58° 08'	12'	7.2
	NGC 7510	23h 11m	+60° 34'	7'	7.9
	Mar 50	23h 15m	+60° 27'	5'	8.5

M 52

NGC 7510

Planetary Nebulae

Con.	Object	R.A.	Dec.	Diam	Mag.
And	NGC 7662	23h 26m	+42° 32′	30″	8.3

Galaxies

Con.	Object	R.A.	Dec.	Diam	Mag.	Type
And	NGC 7640	23h 22m	+40° 51′	11′	10.9	Gx S
Aqr	NGC 7184	22h 03m	-20° 48′	6′	11.4	Gx S
	NGC 7606	23h 19m	-08° 29′	5′	10.9	Gx S
Gru	NGC 7410	22h 55m	-39° 40′	6′	10.6	Gx S
	NGC 7412	22h 56m	-42° 39′	4′	11.2	Gx S
	NGC 7418	22h 57m	-37° 02′	4′	11.2	Gx S
	IC 1459	22h 57m	-36° 28′	3′	10.0	Gx E
	IC 5267	22h 57m	-43° 24′	5′	10.5	Gx S
	NGC 7424	22h 57m	-41° 04′	10′	10.4	Gx S
	NGC 7531	23h 15m	-43° 36′	5′	11.6	Gx S
	NGC 7552	23h 16m	-42° 35′	3′	10.7	Gx S
	NGC 7582	23h 18m	-42° 22′	5′	10.6	Gx S
Peg	NGC 7331	22h 37m	+34° 25′	11′	9.7	Gx S
	NGC 7332	22h 37m	+23° 48′	4′	11.2	Gx S
	NGC 7457	23h 01m	+30° 09′	4′	11.0	Gx S
	NGC 7479	23h 05m	+12° 19′	4′	11.0	Gx S
PsA	NGC 7314	22h 36m	-26° 03′	5′	11.0	Gx S
Scl	NGC 7507	23h 12m	-28° 32′	3′	10.8	Gx E

List 24

Multiple Stars

Con.	Object	R.A.	Dec.	Pair	Sep.	Mag.
And	STF 3050	00h 00m	+33° 43'	AB	2"	6.5, 6.7
Aqr	107 Aqr	23h 46m	-18° 41'	AB	7"	5.7, 6.5
Cas	Sig Cas	23h 59m	+55° 45'	AB	3"	5.0, 7.2
	STF 3053	00h 03m	+66° 06'	AB	15"	6.0, 7.2
Phe	The Phe	23h 40m	-46° 38'	AB	4"	6.5, 7.3
Psc	35 Psc	00h 15m	+08° 49'	AB	12"	6.1, 7.5

Globular Clusters

Con.	Object	R.A.	Dec.	Diam	Mag.
Tuc	NGC 104	00h 24m	-72° 05'	50'	4.0

NGC 104   NGC 55

Galaxies

Con.	Object	R.A.	Dec.	Diam	Mag.	Type
Aqr	NGC 7723	23h 39m	-12° 58'	4'	11.2	Gx S
	NGC 7727	23h 40m	-12° 18'	5'	10.8	Gx S
Cas	IC 10	00h 20m	+59° 18'	6'	10.3	Gx I
Cet	NGC 45	00h 14m	-23° 11'	9'	10.8	Gx S
Peg	NGC 7741	23h 44m	+26° 05'	4'	11.2	Gx S
	NGC 7814	00h 03m	+16° 09'	6'	11.0	Gx S
Phe	IC 5325	23h 39m	-41° 19'	3'	11.0	Gx S
Scl	NGC 7713	23h 36m	-37° 56'	5'	11.2	Gx S
	NGC 7793	23h 58m	-32° 36'	9'	9.2	Gx S
	NGC 55	00h 15m	-39° 13'	32'	8.3	Gx S
	NGC 134	00h 30m	-33° 15'	9'	10.5	Gx S

Appendix

Abbreviations & Acronyms

	Meaning	Relates to	Type
Alp	Alpha	Stars	Designation
Ber	Berkley	Open Clusters	Catalog
Bet	Beta	Stars	Designation
BSO	Brisbane Observatory	Multiple Stars	Catalog
Col	Collinder	Open Clusters	Catalog
Con	Constellation	Objects	Location
Dec	Declination	Objects	Co-ordinates
Del	Delta	Stars	Designation
DSO	Deep Sky Object(s)	All Objects	
DUN	James Dunlop	Multiple Stars	Catalog
Eps	Epsilon	Stars	Designation
Eta	Eta (no abbreviation)	Stars	Designation
Gam	Gamma	Stars	Designation
GC	Globular Cluster	Objects	Object Type
Gx	Galaxies	Objects	Object Type
Gx E	Galaxy - Elliptical	Objects	Object Type
Gx I	Galaxy - Irregular	Objects	Object Type
Gx S	Galaxy – Spiral	Objects	Object Type
H	Herschel	Objects	Object Type
He	Henize	Planetary Neb	Catalog
HJ	John Herschel	Multiple Stars	Catalog
HR	Harvard Revised	Multiple Stars	Catalog
IC	Index Catalog	Misc. Objects	Catalog
Iot	Iota	Stars	Designation
Kap	Kappa	Stars	Designation
King	King	Open Clusters	Catalog
Lam	Lambda	Stars	Designation
LMC	Large Magellanic Cloud	Galaxy	Object Name
M	Messier	Misc. Objects	Catalog
Mar	Markarian	Open Cluster	Catalog
Mel	Melotte	Open Clusters	Catalog
MS	Multiple Star(s)	Objects	Object Type
Mu	Mu (no abbreviation)	Stars	Designation
N/S	North to South	Objects	Direction
Neb	Nebula	Objects	Object Type
NGC	New General Catalog	Misc. Objects	Catalog
Nu	Nu (no abbreviation)	Stars	Designation
OC	Open Cluster	Objects	Object Type
OC/N	Open Cluster with Nebulae	Objects	Object Type
Omi	Omicron	Stars	Designation
Phi	Phi (no abbreviation)	Stars	Designation
Pi	Pi (no abbreviation)	Stars	Designation
Psi	Psi (no abbreviation)	Stars	Designation
R.A.	Right Ascension	Objects	Co-ordinates

Abbreviations & Acronyms (cont.)

	Meaning	Relates to	Type
Rho	Rho (no abbreviation)	Stars	Designation
RMK	Rumker	Multple Stars	Catalog
S	James South	Multiple Stars	Catalog
SHJ	James South / John Herschel	Multiple Stars	Catalog
Sig	Sigma	Stars	Designation
SNR	Supernova Remnant	Objects	Object Type
STF	F. G. W. Struve	Multiple Stars	Catalog
Sto	Stock	Open Clusters	Catalog
STT	O. W. Struve	Multiple Stars	Catalog
Tau	Tau (no abbreviation)	Stars	Designation
The	Theta	Stars	Designation
Tom	Tombaugh	Open Clusters	Catalog
Ups	Upsilon	Stars	Designation
Vel	Vela	Constellation	Name
WNO	US Naval Observatory	Multiple Stars	Catalog
Xi	Xi (no abbreviation)	Stars	Designation
Z	Zenith	Objects	Direction
Zet	Zeta	Stars	Designation

Greek Alphabet

Alp	Alpha	α	**Nu**	Nu	ν
Bet	Beta	β	**Xi**	Xi	ξ
Gam	Gamma	γ	**Omi**	Omicron	ο
Del	Delta	δ	**Pi**	Pi	π
Eps	Epsilon	ε	**Rho**	Rho	ρ
Zet	Zeta	ζ	**Sig**	Sigma	σ
Eta	Eta	η	**Tau**	Tau	τ
The	Theta	θ	**Ups**	Upsilon	υ
Iot	Iota	ι	**Phi**	Phi	φ
Kap	Kappa	κ	**Chi**	Chi	χ
Lam	Lambda	λ	**Psi**	Psi	ψ
Mu	Mu	μ	**Ome**	Omega	ω

Constellations

	Latin	Genitive	Lists
And	Andromeda	Andromedae	1-3, 23-24
Ant	Antila	Antilae	10
Aps	Apus	Apodis	16
Aql	Aquila	Aquilae	19-20
Aqr	Aquarius	Aquarii	21-24
Ara	Ara	Arae	17-18
Ari	Aries	Arietis	2-3
Aur	Auriga	Aurigae	5-7
Boo	Boötes	Boötis	14-15
Cae	Caelum	Caeli	None
Cam	Camelopardalis	Camelopardalis	3-8, 13
Cap	Capricornus	Capricorni	20, 22
Car	Carina	Carinae	7-11
Cas	Cassiopeia	Cassiopeiae	1-3, 23-24
Cen	Centaurus	Centauri	11-15
Cep	Cepheus	Cephei	1, 7, 21-23
Cet	Cetus	Ceti	1-3, 24
Cha	Chamaeleon	Chamaeleontis	None
Cir	Circinus	Circini	15
CMa	Canis Major	Canis Majoris	6-7
CMi	Canis Minor	Canis Minoris	None
Cnc	Cancer	Cancri	8-9
Col	Columba	Columbae	5-6
Com	Coma Berenices	Comae Berenices	12-13
CrA	Corona Australis	Coronae Australis	18-19
CrB	Corona Borealis	Coronae Borealis	16
Crt	Crater	Crateris	11-12
Cru	Crux	Crucis	12-13
Crv	Corvus	Corvi	12-13
CVn	Canes Venatici	Canum Venaticorum	12-14
Cyg	Cygnus	Cygni	19-22
Del	Delphinus	Delphini	21
Dor	Dorado	Doradus	4-6
Dra	Draco	Draconis	12-13, 15-20
Equ	Equuleus	Equulei	21
Eri	Eridanus	Eridani	2-5
For	Fornax	Fornacis	3-4

Constellations (cont.)

	Latin	Genitive	Lists
Gem	Gemini	Geminorum	6-8
Gru	Grus	Gruis	22-23
Her	Hercules	Herculis	16-18
Hor	Horologium	Horologii	3-4
Hya	Hydra	Hydrae	8-15
Hyi	Hydrus	Hydri	3-4
Ind	Indus	Indi	21-22
Lac	Lacerta	Lacertae	22-23
Leo	Leo	Leonis	10-12
Lep	Lepus	Leporis	5-6
Lib	Libra	Librae	15-16
LMi	Leo Minor	Leonis Minoris	9-11
Lup	Lupus	Lupi	14-16
Lyn	Lynx	Lyncis	6-9
Lyr	Lyra	Lyrae	19
Men	Mensa	Mensae	None
Mic	Microscopium	Microscopii	21
Mon	Monoceros	Monocerotis	6-8
Mus	Musca	Muscae	11-14
Nor	Norma	Normae	15-17
Oct	Octans	Octantis	20-22
Oph	Ophiuchus	Ophiuchi	16-18
Ori	Orion	Orionis	5-6
Pav	Pavo	Pavonis	18-21
Peg	Pegasus	Pegasi	21-24
Per	Perseus	Persei	2-5
Phe	Phoenix	Phoenicis	1-2, 24
Pic	Pictor	Pictoris	5-6
PsA	Piscis Austrinus	Piscis Austrini	22-23
Psc	Pisces	Piscium	1-2, 24
Pup	Puppis	Puppis	6-8
Pyx	Pyxis	Pyxidis	9
Ret	Reticulum	Reticuli	3-4
Scl	Sculptor	Sculptoris	1-2, 23-24
Sco	Scorpius	Scorpii	16-18
Sct	Scutum	Scuti	18-19

Constellations (cont.)

	Latin	Genitive	Lists
Ser	Serpens	Serpentis	15-16, 18-19
Sex	Sextans	Sextantis	10-11
Sge	Sagitta	Sagittae	19-20
Sgr	Sagittarius	Sagittarii	18-20
Tau	Taurus	Tauri	4-6
Tel	Telescopium	Telescopii	17-20
TrA	Triangulum Australe	Trianguli Australis	16
Tri	Triangulum	Trianguli	2-3
Tuc	Tucana	Tucanae	1, 22, 24
UMa	Ursa Major	Ursa Majoris	9-14
UMi	Ursa Minor	Ursa Minoris	3, 15, 17
Vel	Vela	Velorum	8-11
Vir	Virgo	Virginis	12-15
Vol	Volans	Volantis	7-8
Vul	Vulpecula	Vulpeculae	19-21

Named Objects

	Designation	Type	Con	List
47 Tucanae	NGC 104	GC	Tuc	1
Acamar	The Eri	MS	Eri	3
Achird	Eta Cas	MS	Cas	1
Adara	Eps CMa	MS	CMa	7
Adhafera	Zet Leo	MS	Leo	10
Albireo	Bet Cyg	MS	Cyg	20
Algieba	Gam Leo	MS	Leo	10
Algiedi	Alp Cap	MS	Cap	20
Alkalurops	Mu Boo	MS	Boo	15
Alkurhah	Xi Cep	MS	Cep	22
Almach	Gam And	MS	And	2
Alnitak	Zet Ori	MS	Ori	6
Alrisha	Alp Psc	MS	Psc	2
Aludra	Eta CMa	MS	CMa	7
Alula Australis	Xi UMa	MS	UMa	11
Andromeda Galaxy	M 31	Gx S	And	1
Antennae Galaxies	NGC 4038/9	Gx S	Crv	12

Named Objects (cont.)

	Designation	Type	Con	List
Aristotle's Cluster	M 41	OC	CMa	7
Baade's Window	NGC 6522	GC	Sgr	18
Barnard's Galaxy	NGC 6822	Gx I	Sgr	20
Beehive Cluster	M 44	OC	Cnc	9
Big Bear Pair	NGC 3718/29	Gx S	UMa	12
Black Eye Galaxy	M 64	Gx S	Com	13
Blinking Planetary	NGC 6826	PN	Cyg	20
Blue Planetary	NGC 3918	PN	Cen	12
Blue Puff Nebula	NGC 1514	PN	Tau	4
Blue Snowball Nebula	NGC 7662	PN	And	23
Blue Spiral Galaxy	NGC 1232	Gc S	Eri	3
Bode's Galaxy	M 81	Gx S	UMa	10
Box Galaxy	NGC 4449	Gx I	CVn	12
Bubble Nebula	NGC 246	PN	Cet	1
Butterfly Cluster	M 6	OC	Sco	18
Butterfly Galaxies	NGC 4567/8	Gx S	Vir	13
California Nebula	NGC 1499	Neb	Per	4
Caroline's Cluster	M 48	OC	Hya	8
Castor	Alp Gem	MS	Gem	8
Cat's Eye Nebula	NGC 6543	PN	Dra	18
Centaurus A	NGC 5128	Gx E	Cen	13
Cetus A	M 77	Gx S	Cet	3
Cetus Bubble Nebula	NGC 246	PN	Cet	1
Christmas Tree Cluster	NGC 2264	OC/N	Mon	7
Cigar Galaxy	M 82	Gx I	UMa	10
Cleopatra's Eye	NGC 1535	PN	Eri	4
Clown Face Nebula	NGC 2392	PN	Gem	7
Coalsack Cluster	NGC 4609	OC	Cru	13
Cocoon Galaxy	NGC 4490/85	Gx S	CVn	13
Cocoon Nebula	IC 5146	OC/N	Cyg	22
Coma Star Cluster	Mel 111	OC	Com	12
Cone Nebula	NGC 2264	OC/N	Mon	7
Cor Caroli	Alp CVn	MS	CVn	13
Crab Nebula	M 1	SNR	Tau	6
Crescent Nebula	NGC 6888	Neb	Cyg	20
Dabih	Bet Cap	MS	Cap	20
Denning's Galaxy	IC 342	Gx S	Cam	4
Discus Galaxy	NGC 4216	Gx S	Vir	12
Double Cluster	NGC 869/889	OC	Per	2
Double Double	Eps Lyr	MS	Lyr	19
Dubhe	Alp UMa	MS	UMa	11
Dumbbell Nebula	M 27	PN	Vul	20

Named Objects (cont.)

	Designation	Type	Con	List
E.T. Cluster	NGC 457	OC	Cas	1
Eagle Nebula	M 16	OC/N	Ser	18
Eight Burst Nebula	NGC 3132	PN	Vel	10
Eskimo Nebula	NGC 2392	PN	Gem	7
Eta Carinae Nebula	NGC 3372	Neb	Car	11
Exploding Galaxy	M 82	Gx I	UMa	10
Fireworks Galaxy	NGC 6946	Gx S	Cyg	21
Flaming Star Nebula	IC 405	Neb	Aur	5
Flocculent Spiral Galaxy	NGC 4414	Gx S	Com	12
Fornacis	Alp For	MS	For	3
Frisbee Galaxy	NGC 4274	Gx S	Com	12
Gabriela Mistral Nebula	NGC 3324	OC/N	Car	11
Gem Cluster	NGC 3293	OC	Car	11
Ghost of Jupiter	NGC 3242	PN	Hya	10
Gold Dust Cluster	M 37	OC	Aur	6
Golf Ball Cluster	NGC 752	OC	And	2
Graffias	Bet Sco	MS	Sco	16
Grand Design Galaxy	M 74	Gx S	Psc	2
Great Hercules Cluster	M 13	GC	Her	17
Great Orion Nebula	M 42	Neb	Ori	6
Grus Quartet	NGC 7582	Gx S	Gru	23
Hamburger Galaxy	NGC 3628	Gx S	Leo	11
Heart Nebula	IC 1805	OC/N	Cas	3
Helix Nebula	NGC 7293	PN	Aqr	22
Hockey Stick Galaxy	NGC 4656	Gx S	CVn	13
Horseshoe Cluster	NGC 663	OC	OC	2
Horseshoe Nebula	M 17	OC/N	Sgr	18
Hyades	Mel 25	OC	Tau	4
Jabbah	Nu Sco	MS	Sco	16
Jewel Box	NGC 4755	OC	Cru	13
Kaffaljidhma	Gam Cet	MS	Cet	3
Keystone Cluster	M 13	GC	Her	17
Kuma	Nu Dra	MS	Dra	18
Lagoon Nebula	M 8	OC/N	Sgr	18
Lambda Centauri Nebula	IC 2944	OC/N	Cen	12
Large Magellanic Cloud		Gx S	Dor	5
Little Dumbbell Nebula	M 76	PN	Per	2
Little Gem Nebula	NGC 6818	PN	Sgr	20

Named Objects (cont.)

	Designation	Type	Con	List
Marfik	Kap Her	MS	Her	16
Mechain's Galaxy	M 106	Gx S	CVn	12
Meissa	Lam Ori	MS	Ori	6
Mesarthim	Gam Ari	MS	Ari	2
Mintaka	Del Ori	MS	Ori	6
Mirror Ball Cluster	M 10	GC	Oph	17
Mizar & Alcor	Zet UMa	MS	UMa	13
Nair al Saif	Iot Ori	MS	Ori	6
North American Nebula	NGC 7000	Neb	Cyg	21
Omega Cluster	NGC 5139	GC	Cen	13
Omega Nebula	M 17	OC/N	Sgr	18
Omicron Vela Cluster	IC 2391	OC	Vel	9
Orb Cluster	M 92	GC	Her	17
Oriani's Galaxy	M 61	Gx S	Vir	12
Orion Nebula	M 42	Neb	Ori	6
Owl Cluster	NGC 457	OC	Cas	1
Owl Nebula	M 97	PN	UMa	11
Pac Man Nebula	NGC 281	OC/N	Cas	1
Pavo Cluster	NGC 6752	GC	Pav	19
Pearl Cluster	NGC 3766	OC	Cen	12
Pelican Nebula	IC 5070	Neb	Cyg	21
Pinwheel Galaxy	M 101	Gx S	UMa	14
Pleiades	M 45	OC/N	Tau	4
Polaris	Alp UMi	MS	UMi	3
Praesepe	M 44	OC	Cnc	9
Propeller Galaxy	NGC 4536	Gx S	Vir	13
Ptolemy's Cluster	M 7	OC	Sco	18
Rasalgethi	Alp Her	MS	Her	17
Retina Nebula	IC 4406	PN	Lup	14
Ring Nebula	M 57	PN	Lyr	19
Robin's Egg Nebula	NGC 1360	PN	For	4
Rosette Nebula	NGC 2244	OC/N	Mon	7
Running Man Cluster	NGC 2516	OC	Car	8
Running Man Nebula	NGC 1977	OC/N	Ori	6
S Normae Cluster	NGC 6087	OC	Nor	16
Sagittarius Star Cloud	M 24	OC	Sgr	18
Sailboat Cluster	NGC 225	OC	Cas	1
Salt & Pepper Cluster	M 52	OC	Cas	23
Saturn Nebula	NGC 7009	PN	Aqr	21
Sculptor Spiral Galaxy	NGC 300	Gx S	Scl	1
Seven Sisters	M 45	OC/N	Tau	4

	Designation	Type	Con	List
Sheliak	Bet Lyr	MS	Lyr	19
Silver Coin Galaxy	NGC 253	Gx S	Scl	1
Silver Needle Galaxy	NGC 4244	Gx S	CVn	12
Small Megallanic Cloud	NGC 292	Gx S	Tuc	1
Snowball Cluster	M 12	GC	Oph	17
Sombrero Galaxy	M 104	Gx S	Vir	13
Southern Pinwheel	M 83	Gx S	Hya	14
Southern Pleiades	IC 2602	OC	Car	11
Spider Web Galaxy	NGC 3521	Gx S	Leo	11
Spindle Galaxy	M 102	Gx S	Dra	15
Spiral Nebula	NGC 5189	PN	Mus	14
Splinter Galaxy	NGC 5907	Gx S	Dra	15
Star Queen Nebula	M 16	OC/N	Ser	18
Starfish Cluster	M 93	OC	Pup	8
Subaru	M 45	OC/N	Tau	4
Suhail al Muhif	Gam Vel	MS	Vel	8
Sunflower Galaxy	M 63	Gx S	Com	13
Swan Nebula	M 17	OC/N	Sgr	18
Tao Galaxy	NGC 4535	Gx S	Vir	13
Tarantula Nebula	NGC 2070	Neb	Dor	6
Tegmen	Zet Cnc	MS	Cnc	8
The Eyes	NGC 4435/38	Gx S	Vir	12
Topsy Turvy Galaxy	NGC 1313	Gx S	Ret	3
Trapezium	The Ori	MS	Ori	6
Triangulum Galaxy	M 33	Gx S	Tri	2
Trifid Nebula	M 20	Neb	Sgr	18
UFO Galaxy	NGC 2683	Gx S	Lyn	9
Veil Nebula (East)	NGC 6992	SNR	Cyg	21
Velvet Cluster	M 46	OC	Pup	8
Virgo Cluster	NGC 5634	GC	Vir	14
Vortex Galaxy	NGC 3198	Gx S	UMa	10
Whale Galaxy	NGC 4631	Gx S	CVn	13
Whirlpool Galaxy	M 51	Gx S	CVn	13
Wild Duck Cluster	M 11	OC	Sct	19
Wishing Well Cluster	NGC 3532	OC	Car	11
X Cluster	M 34	OC	Per	3
X Ray Galaxy	NGC 1097	Gx S	For	3
Zubenelgenubi	Alp Lib	MS	Lib	15

Latitude & Declination Table

This table shows the declination ranges of objects that will appear from various latitudes across the world. It should be remembered that although an object may be theoretically observable from a location, its visibility will greatly depend upon a number of factors, including its brightness and the airmass (ie, the thickness of the atmosphere) its light has to travel through.

At the zenith (directly overhead) the atmosphere is at its thinnest and the airmass is said to be one. The object is therefore at its best visibility. However, the atmosphere is thicker closer to the horizon and this might distort the view or render some fainter objects invisible.

For example, at thirty degrees above the horizon objects are observed through an atmosphere of two airmasses, or twice the density of the atmosphere at the zenith. Objects at an altitude of twenty degrees are observed through an atmosphere of three airmasses.

Objects that appear lower than thirty degrees are still observable; however, as their light is travelling through a thicker portion of the atmosphere (for example, five or six airmasses at an altitude of ten degrees) they may be difficult to observe or not visible at all. For example, a multiple star with a wide separation or a bright open cluster may be visible, but a galaxy or planetary nebula might not.

Geographical Latitude	Visible Object Range	Objects Above 20° (3x Airmass)	Objects Above 30° (2x Airmass)
70° N	-20° to +90°	+00° to +90°	+10° to +90°
60° N	-30° to +90°	-10° to +90°	+00° to +90°
50° N	-40° to +90°	-20° to +90°	-10° to +90°
40° N	-50° to +90°	-30° to +90°	-20° to +90°
30° N	-60° to +90°	-40° to +90°	-30° to +90°
20° N	-70° to +90°	-50° to +90°	-40° to +80°
10° N	-80° to +90°	-60° to +80°	-50° to +70°
0° N	-90° to +90°	-70° to +70°	-60° to +60°
10° S	-90° to +80°	-80° to +60°	-70° to +50°
20° S	-90° to +70°	-90° to +50°	-80° to +40°
30° S	-90° to +60°	-90° to +40°	-90° to +30°
40° S	-90° to +50°	-90° to +30°	-90° to +20°
50° S	-90° to +40°	-90° to +20°	-90° to +10°
60° S	-90° to +30°	-90° to +10°	-90° to +00°
70° S	-90° to +20°	-90° to +00°	-90° to -10°

Recommended Resources

Books

- *Astronomy Hacks* – Robert Bruce Thompson & Barbara Fritchman Thompson
- *Binocular Highlights* – Gary Seronik
- *Celestial Harvest* – James Mullaney
- *Celestial Sampler* – Sue French
- *Cosmic Challenge* – Philip S. Harrington
- *Deep-Sky Wonders* – Sue French
- *Double Stars for Small Telescopes* – Sissy Haas
- *Illustrated Guide to Astronomical Wonders* – Robert Bruce Thompson & Barbara Fritchman Thompson
- *Observer's Sky Atlas, The* – E. Karkoschka
- *Pocket Sky Atlas* – Roger W. Sinnott
- *Star Names – Their Lore and Meaning* – Richard Hinckley Allen
- *Star-Hopping for Backyard Astronomers* – Alan M. MacRobert
- *Turn Left at Orion* – Guy Consolmagno and Dan M. Davis

Software

- *Mobile Observatory* by Wolfgang Zima (http://zima.co.)
- *Sky Tools* by Greg Crinklaw (http://www.skyhound.com)
- *Stellarium* (Open Source software – http://www.stellarium.org)

Facebook Groups

- *Astronomy for Beginners (A4B)*
- *Astronomy for Fun*
- *Astronomy Workfile*
- *Online Astronomy Society*
- *Space Science and Astronomy for Home Educators*
- *Telescope Addicts – Astronomy & Astrophotography*
- *UK Astronomy*